SUSTAINABLE HOME DESIGN BY APPLYING CONTROL SCIENCE

Authored by **Kazutoshi Fujihira**

Sustainable Home Design by Applying Control Science
http://dx.doi.org/10.5772/64660
Authored by Kazutoshi Fujihira

Contributors

Kazutoshi Fujihira

Published by InTech
Janeza Trdine 9, 51000 Rijeka, Croatia

Publishing Process Manager Romina Rovan
Technical Editor SPi Global
Cover InTech Design team

First published November, 2017
Printed in Croatia
Legal deposit, Croatia: National and University Library in Zagreb

Additional hard copies can be obtained from orders@intechopen.com

Sustainable Home Design by Applying Control Science, Authored by Kazutoshi Fujihira
p. cm.
Print ISBN 978-953-51-3657-6
Online ISBN 978-953-51-3658-3

For EU product safety concerns:
IN TECH d.o.o., Prolaz Marije Krucifikse Kozulić 3, 51000 Rijeka, Croatia,
info@intechopen.com or visit our website at intechopen.com.

Meet the author

Kazutoshi Fujihira, the founder and head of *Institute of Environmentology*, is an innovative scientist in the fields of environment and sustainability. His research started with clarifying causal relation between humans and environmental problems. After that, applying control science, he has been developing measures to control human activities towards sustainability. Recently, synthesizing these studies and his architectural background, he has produced a methodology for promoting sustainable home design. He is the author of several books, including *Introduction to Environmentology* and *Complex Systems, Sustainability and Innovation* and numerous research papers. He is also the recipient of the *2015 Japan Wooden Houses and Industry Association President Award* and *2017 Albert Nelson Marquis Lifetime Achievement Award*.

Contents

Preface

Homes are the places where people spend the largest part of their time in a day. Many older people, in particular, spend most of their time in their homes. Meanwhile, homes are used for a very long time. Homes which are built now are expected to be used at least until the end of the twenty-first century.

On the other hand, our surrounding environment and circumstances are noticeably changing. It is regarded as certain that especially two major changes progress globally during the twenty-first century. They are "climate change" and "ageing population." People in the world have to not only curb the progress of climate change but also reduce adverse effects caused by climate change. The other global change, ageing population, is a serious threat to public finances through the increase in medical and nursing care expenses.

Facing these changes, housing also has to be changed. Homes need to be transformed into those which contribute to curbing the progress of climate change and reducing harmful influences caused by climate change, as well as reducing medical and nursing care expenses in ageing population. How homes should be designed so as to comprehensively meet these various needs? How such design can be promoted in the housing market? This monograph will answer these questions by utilizing a new methodology, namely, the application of "control" science.

This monograph consists of six chapters. Chapter 1 clarifies major sustainability issues related to housing, mainly focusing on climate change and ageing population. Chapter 2 outlines existing Japanese systems related to sustainable housing and clarifies the remaining issues. The third chapter demonstrates the basic schemes, that is, the "basic control system for sustainability", "model of sustainability" and "two-step preparatory work for sustainable design". Based on these basic schemes, Chapter 4 presents the methodology, namely, the "control system for promoting sustainable home design" with the "sustainable design guidelines" and "sustainability checklist". Chapter 5 illustrates a case study, in which, following the methodology, we designed a home and actually constructed it. The last chapter discusses the effectiveness, characteristics and future prospects of the methodology.

The author of this monograph has continued studies on environment and sustainability for a long time, presiding over the Institute of Environmentology, Tokyo. Recently, he has been tackling sustainability issues, applying control science. Meanwhile, he has engaged in planning public buildings and supervising construction at a local government for nearly 20 years. He also has experiences of designing detached houses. Synthesizing these studies and experiences, he has produced this monograph.

More directly, the author has developed his ideas from the two already published book chapters into this monograph. The two book chapters have been "An Approach to Sustainable Development by Applying Control Science"in "Sustainable Development – Policy and Urban Development – Tourism, Life Science, Management and Environment" and "System Control for Sustainability: Application to Building Design" in "Complex Systems, Sustainability and Innovation", which are equivalent to the prototype and the summary of this monograph, respectively. Both publications have been published by InTechOpen, as well as this research book.

I would like to express my gratitude to the staff of InTechOpen Publisher for their great efforts in bringing the book to fruition. Especially, I would like to convey my appreciation to the Publishing Process Manager, Ms. Romina Rovan, for her great efforts in coordinating the publishing process. Moreover, I would also like to acknowledge the reviewer for his valuable and constructive comments on the manuscript.

I would like to express my gratitude to Mr. Vance Carothers for his valuable advice and suggestions on the English and diagram expressions of this monograph. I would like to acknowledge the staff of IK Home Inc., or Eyeful Home Setagaya Branch, for their careful work in designing and building the home in the case study. Lastly, I would also like to thank Kigokoro Plants and Gardenworks and Eco-gardener Yoshiko, for their skilful landscape gardening design and practice in the case study.

Kazutoshi Fujihira
Institute of Environmentology,
Tokyo, Japan

Chapter 1

Requirements for Sustainable Housing Design

Kazutoshi Fujihira

Additional information is available at the end of the chapter

http://dx.doi.org/10.5772/intechopen.71322

Abstract

Buildings including homes are related to a variety of environmental, social, and economic problems. In the twenty-first century, especially two major problems, namely "climate change" and financial problems resulting from "aging population," are expected to become more serious globally. In order to curb the progress of climate change, building sector has to strengthen "mitigation" measures, including improvement in energy efficiency of buildings and utilization of renewable energy. Building sector must also strengthen "adaptation" measures, aiming to reduce adverse effects caused by climate change. On the other hand, homes need to be transformed into those which contribute to reduce illnesses and injuries, which tend to increase in accordance with aging population. Taking accessible and universal design into homes is effective to increase mobility of occupants as well as to prevent injuries. Homes are used for a very long time; therefore, these considerations need to be comprehensively taken into homes from the beginning.

Keywords: climate change, mitigation, adaptation, aging population, accessibility, universal design

1. Introduction

Buildings including homes are the places where people spend the largest part of their time. According to a survey in the U.S., people spend 87% of their time in enclosed buildings, made up of 69% in residences and 18% in other types of buildings [1]. Meanwhile, buildings including homes are used for a very long time. Longer lifespan of buildings is desirable, since it leads to reducing natural resources for construction as well as waste.

Sheltering occupants from severe climate and weather, homes are expected to protect occupants' health, safety, and well-being. In the twenty-first century, two major significant changes progress inside and outside houses. They are "climate change" and demographic change, namely "aging population". That is to say, it is regarded as certain that human activities influence climate and increase extreme weather events, while humans themselves increase in average age.

Facing these changes, homes also have to be changed toward sustainability. Focusing mainly on climate change and aging population, this chapter describes main points that the two major changes require homes to deal with, so as to lead to sustainability.

2. Climate change and buildings

2.1. Climate change: observed changes and their causes

Greenhouse gas (GHG) emissions originating in human activities have grown, and recent global GHG emissions are the highest in the history. According to the Intergovernmental Panel on Climate Change (IPCC), GHG emissions from the building sector have more than doubled since 1970 to reach 9.18 $GtCO_2$-equivalent in 2010, representing 19% of all global 2010 GHG emissions. In 2010, the building sector accounted for 32% (24% for residential and 8% for commercial) of total global final energy use and 51% of global electricity consumption [2].

Human influence on the climate system is clear. Atmospheric concentrations of greenhouse gases such as carbon dioxide (CO_2), methane (CH_4), and nitrous oxide (N_2O) have obviously increased. In recent decades, changes in climate have caused impacts on natural and human systems on all continents and across the oceans. Impacts from recent climate-related extremes, such as heat waves, droughts, floods, cyclones, and wildfires, reveal significant vulnerability and exposure of some ecosystems and many human systems to current climate variability (very high confidence) [3].

It is very likely that the number of cold days and nights has decreased and the number of warm days and nights has increased on the global scale. It is likely that the frequency of heat waves has increased in large parts of Europe, Asia, and Australia [3]. Extreme heat events currently result in increases in mortality and morbidity in North America (very high confidence) and in Europe with impacts that vary according to people's age, location, and socio-economic factors (high confidence) [3]. Also, in recent Japan, there has been an increasing trend in the number of deaths from heat stroke (**Figure 1**); experts point out that a cause of this increasing trend is recent rising tendency in temperature [4, 5].

It is likely that extreme sea levels have increased since 1970, being mainly the result of mean sea level rise. There are likely more land regions where the number of heavy precipitation events has increased than where it has decreased [3]. It is virtually certain that intense tropical cyclone activity has increased in the North Atlantic, since 1970 [3]. Similarly, according to a latest study, typhoons that strike East and Southeast Asia have intensified by 12–15% since the late 1970s [6].

2.2. Warming versus cumulative CO_2 emissions

The Fifth Assessment Report of the IPCC has concluded that "cumulative emissions of CO_2 largely determine global mean surface warming by the late 21st century and beyond" [3].

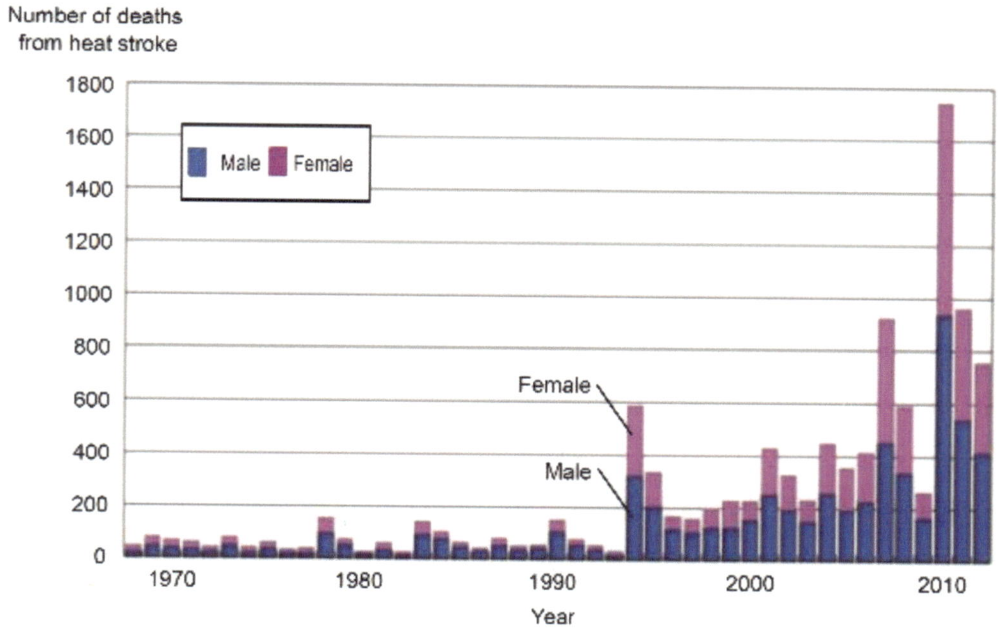

Figure 1. Changes in the number of annual deaths from heat stroke by sex in Japan (1968–2012) [4]. Source: Ministry of Health, Labour and Welfare, Vital Statistics.

"Cumulative emissions of CO_2" means the total amount of carbon dioxide that human beings have emitted into the atmosphere [7].

Figure 2 demonstrates the relationship between "cumulative anthropogenic CO_2 emissions from 1870" and "temperature change relative to 1861–1880". The colored plume illustrates the multimodel spread over the IPCC's latest four scenarios named representative concentration pathways. Meanwhile, the hollow ellipses show total anthropogenic warming in 2100 versus cumulative CO_2 emissions from 1870 to 2100. The "baselines" means scenarios without additional efforts to reduce GHG emissions beyond those in place today. The ellipses with figures show the scenario categories of the IPCC's Working Group III, according to levels of efforts to reduce GHG emissions; figures such as "530–580" mean projected CO_2-equivalent concentration levels (in ppm) in 2100. The black-filled ellipse represents observed emissions to 2005 and observed temperatures in the decade 2000–2009 with associated uncertainties. In short, both the plume and ellipses in **Figure 2** indicate a strong, consistent, and almost linear relationship between cumulative CO_2 emissions and projected global temperature change to the year 2100 [3].

Figure 2 also shows that "regardless of the scale of reducing GHG emissions, cumulative CO_2 emissions will increase, and therefore, global surface temperature is projected to rise over the 21st century" [3, 8]. As a result, the adverse impacts of projected climate change, such as heat waves, floods, cyclones, and wildfires, will increasingly become more severe.

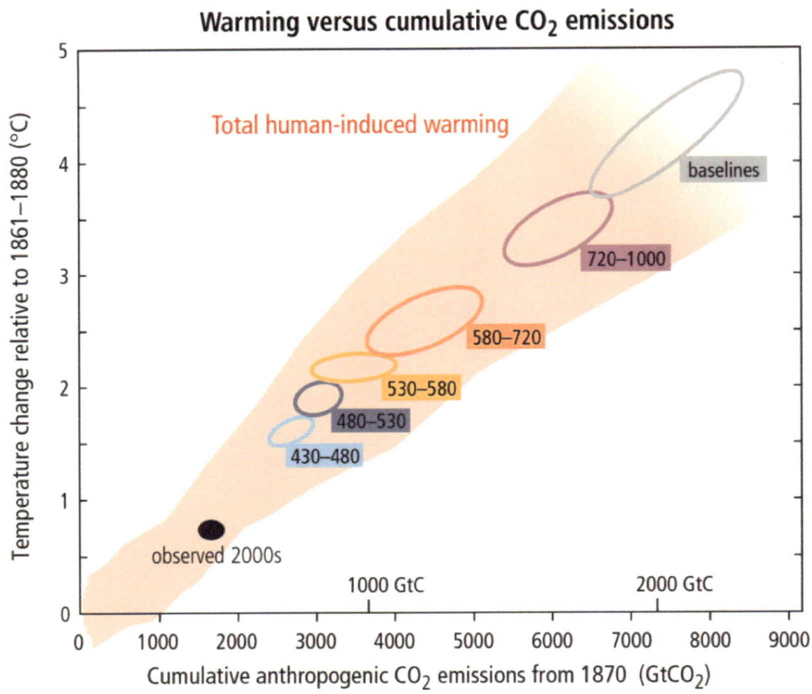

Figure 2. Relationship between cumulative CO_2 emissions and projected global temperature change to the year 2100 [3]. Notes: (1) Colored plume shows the spread of past and future projections from a hierarchy of climate-carbon cycle models driven by historical emissions and the four representative concentration pathways over all times out to 2100 and fades with the decreasing number of available models. (2) The hollow ellipses show total anthropogenic warming in 2100 versus cumulative CO_2 emissions from 1870 to 2100 from a simple climate model (median climate response) under the scenario categories used in Working Group III. The width of the ellipses in terms of temperature is caused by the impact of different scenarios for non-CO_2 climate drivers. (3) The filled black ellipse shows observed emissions to 2005 and observed temperatures in the decade 2000–2009 with associated uncertainties.

2.3. Mitigation, adaptation, and sustainable development

Reducing climate change risks requires "adaptation" as well as "mitigation" (**Table 1**) [3, 8]. The IPCC defines "mitigation" as "a human intervention to reduce the sources or enhance the sinks of greenhouse gases" [3]. Mitigation measures include energy conservation, harnessing renewable energy, and absorbing CO_2 through afforestation. On the other hand, "adaptation" is defined as "the process of adjustment to actual or expected climate and its effects" [3]. Adaptation seeks to moderate or avoid harmful influences caused by climate change [3]. Adaptation strategies include heatstroke prevention, measures against floods, cyclones, drought, infectious diseases, and high-temperature damage to cultivated crops.

Neither adaptation nor mitigation alone can avoid all climate change impacts [7]. Adaptation and mitigation are complementary strategies for reducing and managing the risks of climate change [3].

The chapter of "Climate-Resilient Pathways: Adaptation, Mitigation, and Sustainable Development" in *Climate Change 2014* explains, "It takes sustainable development as the ultimate goal and

http://dx.doi.org/10.5772/intechopen.71322

Strategy	Definition	Typical examples
Mitigation	Human intervention to reduce the sources or enhance the sinks of greenhouse gases	• Energy conservation • Harnessing renewable energy • Conserving timber resources • Carbon dioxide capture and storage
Adaptation	Process of adjustment to actual or expected climate and its effects	• Measures against inland and coastal floods • Measures for dealing with water shortages • Heatstroke warning systems • Building resilient infrastructure

Table 1. Definition and typical examples of mitigation and adaptation.

considers mitigation as a way to keep climate change moderate rather than extreme. Adaptation is considered a response strategy to anticipate and cope with impacts that cannot be (or are not) avoided under different scenarios of climate change" [9].

2.4. Mitigation and adaptation strategies in building sector

There are a variety of mitigation and adaptation options in the building sector as well as other sectors.

Buildings represent a critical piece of a low-carbon future and a global challenge for integration with sustainable development (robust evidence and high agreement) [2]. Existing and future buildings will determine a large proportion of global energy demand. Current trends indicate the potential for massive increases in energy demand and associated emissions. However, if today's cost-effective best practices and technology are broadly diffused, final energy may stay constant or even decline by mid-century, as compared to today's levels [2].

The IPCC's Fifth Assessment Report has classified building sector's mitigation options by strategies: improvement of "carbon efficiency", "energy efficiency of technology", and "system efficiency", as well as "service demand reduction". **Table 2** shows extracts from such mitigation options of these four strategies [2].

In addition to mitigation strategies, adapting building designs for climate change is about dealing with the unavoidable [10]. There is no consensus on definitions of climate adaptive buildings [2]. However, there is growing awareness that design practices need to take into account the predictions of increased risk and intensity of extreme events [10].

Table 3 presents main predicted risks and examples of adapting building/housing design options [10, 11]. Strategies to reduce climate-related risks differ across regions [3]. Therefore, people involved in promoting sustainable design and designing sustainable buildings in each region need to appropriately predict future risks and plan effective strategies.

Mitigation strategy	Major mitigation options of buildings
Carbon efficiency	• Building integrated renewable energy system • Fuel switching to low-carbon fuels
Energy efficiency of technology	• High-performance building envelope • Efficient appliances and lighting • Efficient heating, ventilation, and air-conditioning systems
System efficiency	• Passive house standard • Net zero energy and energy plus buildings
Service demand reduction	• Behavioral change • Lifestyle change

Table 2. Major mitigation options of buildings by strategies.

Extreme weather events and impacts	Examples of adaptive strategies for building/housing design
Increasing temperatures	• High thermal insulation • External shading • Natural ventilation • Green roof
Cyclones and storms	• More wind-resistant buildings • Select impact-resistant exterior materials • Install window protection
High intense rainfall, flooding, and rising sea levels	• Use water-resistant materials • Raise buildings off the ground
Low rainfall and drought	• Use water-saving fixtures and appliances • Harvest rainwater
Wildfire	• Use fire-resistant exterior materials

Table 3. Examples of adaptive strategies for building/housing design by impacts.

3. Aging population and housing

3.1. Aging population: fierce threat to public finances

While growing human activities are aggravating climate change and seriously harming environmental sustainability, "aging population" is becoming a fierce challenge to social and economic sustainability. According to the book titled *Agequake*, in the twenty-first century, there will be more people older than younger ones. The "agequake" will turn the age pyramids upside down, first in Japan and Europe, but ultimately throughout the world [12]. Europe is projected to remain the most aged region in the coming decades, with 34% of the population aged 60 years or over in 2050 [13]. The European Union warns that the aging of the population is becoming a growing challenge to the sustainability of public finances. The increase of the ratio between the number of retirees and the number of workers will amplify expenditure on public pensions and health and long-term care and thus put a burden on maintaining a sound balance between future public expenditure and tax revenues [14]. The demographic change of Japan is more drastic (**Figure 3**); Japan's population aged 65 years or over is projected to reach about 40% in 2060 [15]. Similarly, demographic pressures will continue to mount and add to concerns about fiscal sustainability in Japan [16].

3.2. Links between unsuitable housing and ill health

Ed Harding: International Longevity Centre UK clearly states, "Suitable housing is central to the challenge of population ageing" [17]. In the UK, older people spend between 70 and 90% of their time in their homes, much more than any other age group [17]. A survey in Japan also indicates similar results; people aged 70 years or over spend on average about 83% of their time in their houses, which is considerably larger than any other age group [18]. Unsuitable housing has direct and proven linkages with ill health, including allergy, asthma, heart disease, hypertension, falls, and annoyance [17, 19, 20].

For example, in Japan, illnesses and deaths of elderly people are related to lower thermal performance of homes in both winter and summer. In winter, many people die by drowning in homes' bath. The number of sudden bath-related deaths increased to 4866 in 2014 [21], which exceeded the number of traffic deaths in the same year, namely 4113 [22]. Approximately 90% of the bath-related deaths are the elderly, people aged 65 years or over [21]. Monthly data on

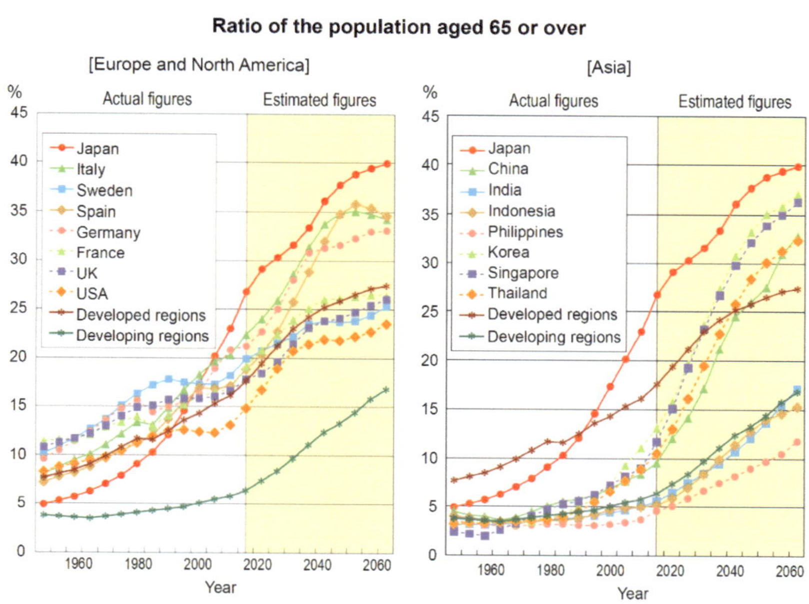

Figure 3. Ratio of the population aged 65 years or more [15]. Source: UN, World Population Prospects: The 2015 Revision. However, data on Japan are taken before 2010 from "Population Census" (Ministry of Internal Affairs Communications), 2015 from the estimated population based on the "2015 Population Census," and after 2020 from "Population Projections for Japan (January 2012 Estimates)" based on the estimated figure with Medium-Fertility and Medium-Mortality Assumption (National Institute of Population and Social Security Research).

such deaths demonstrate that as outdoor temperatures become lower, the more deaths occur [21, 23]. An important reason for frequent bath-related deaths in winter is poor thermal insulation performance of homes. In many cases, elderly people move from a locally heated living room to a cold changing room and bathe in hot water. As a result, exposed to sudden temperature changes, they have a stroke, myocardial fraction or faint, and finally, drown [21, 23].

Meanwhile, in summer, the recent increases in heat stroke patients and deaths (**Figure 1**) are also related to aging population as well as rise in temperature [4, 5]. The ratios of the elderly to the total heat stroke patients and deaths have been increasing. For example, the ratio of the elderly to the total heat stroke deaths rose from 54% in 1995, to 72% in 2008, and 79% in 2010 [4]. At the same time, the number of elderly people having heat stroke in their "homes" has also been growing. Heat stroke cases in "homes" account for over 50% of the total heat stroke cases of elderly in recent years [4]. Surveys in Japan have indicated that countermeasures against heat stroke include improvement in housing thermal performance, such as thermal insulation performance and sunlight adjustment capability [24].

3.3. Fall-and-slip–related injuries in homes

In addition to thermal performance–related illnesses and deaths, fall-and-slip–related injuries in homes are also a serious health problem. In Australia, slips, trips, and falls in buildings constitute a large and costly public health problem, which is expected to grow due to aging of the Australian population. The most common place of occurrence has been in the "home", which accounts for 62% of all fall-and-slip–based injuries, according to a survey conducted from 2002/02 to 2004/05. The total hospital cost for such home injuries in this period has been estimated at about 1.8 billion A$ [25].

Also in Japan, slips and falls in homes is a major problem which requires a large amount of medical costs. **Table 4** shows the extract of the data on accidents involving injuries of adults that Consumer Affairs Agency, Government of Japan and National Consumer Affairs Center of Japan, have collected from 13 medical institutions. The place where accidents occur most frequently is the "home." The ratios of accidents occurring in homes to all accidents have exceeded 70% in all of the three age groups. Next, breaking down the data by causes of accidents, we realize that the ratios of accidents caused by "falls" and "slips/trips" increase with age. Moreover, the ratios of "middle-level injury" and "serious/critical injury and death" are higher in elderly people, especially in the group of "75 years old or over" [26]. In addition, "middle-level injury" is the severity of injuries, which hospitalization is necessary, but the injury is not life-threatening. These figures show a tendency of an increase in the degree of severity in accordance with aging.

3.4. Need for accessibility and universal design in homes

In order to prevent accidents involving injuries, homes need to improve safety. At the same time, aging population requires homes to have accessibility to disabled people. For instance, a research paper titled *Aging and Disability: Implications for the Housing Industry and Housing Policy in the United States* clearly raises a problem. "Since disability rates increase with age,

Age group		20–64	65–74	75+
Ratio of accidents involving injuries occurring in homes to all accidents		71.4%	72.4%	79.8%
Ratios of accidents caused by falls and slips/trips to all causes of accidents involving injuries in homes	Falls	17.8%	30.1%	30.6%
	Trips/slips	7.1%	18.2%	29.1%
	(Subtotal)	(24.9%)	(48.3%)	(59.7%)
Degree of severity	Slight injury	78.3%	64.2%	55.9%
	Middle-level injury	20.1%	31.3%	35.6%
	Serious/critical injury & death	1.6%	4.5%	8.5%

Note: "Slight injury" is the severity of injuries that does not require hospitalization. "Middle-level injury" is the severity of injuries that hospitalization is necessary but the injury is not life-threatening.
Source: National Consumer Affairs Center of Japan, 2013.

Table 4. Accidents involving injuries occurring in homes by age.

population aging will bring substantial increases in the number of disabled persons and have a significant impact on the nation's housing needs" [27]. In fact, disability rates rise in accordance with age, as shown in **Table 5**. "DIS-1" on the left side of this table means "individuals with physical limitations," which is defined as persons with long-lasting conditions that substantially limit one or more physical activities such as walking, climbing stairs, lifting, and carrying. On the other hand, "DIS-2" stands for "individuals with self-care limitations", which is defined as persons with conditions lasting 6 months or more that make it difficult to dress, bathe, or get around inside the home [27].

Furthermore, aging population will inevitably increase the number of households with at least one disabled resident over the next several decades. This study also estimates that there is a 60% probability that a newly built single-family detached unit will house at least

[DIS-1] Individuals with physical limitations			[DIS-2] Individuals with self-care limitations		
Age	**Male (%)**	**Female (%)**	**Age**	**Male (%)**	**Female (%)**
<35	1.8	1.7	<35	1.0	0.8
35–44	5.5	5.6	35–44	1.6	1.7
45–54	8.9	9.5	45–54	2.4	2.7
55–64	15.7	16.2	55–64	3.7	4.2
65–74	21.8	23.2	65–74	5.6	6.7
75–84	31.3	36.4	75–84	11.3	15.2
85+	47.3	60.8	85+	24.7	37.7

Note: Data are for the population aged 5 and older.

Table 5. Individual disability rates by age and sex in the US in 2000 [27].

one "DIS-1" disabled resident during its expected lifetime and a 25% probability of housing at least one "DIS-2" disabled resident. The authors of the study recommend that planners should broaden their vision of the built environment to include the accessibility of the housing stock [27].

Incorporating "universal design" features into homes and improving accessibility brings various benefits. Universal design is defined as the design of products, environments, programs, and services to be usable by all people to the greatest extent possible, without the need for adaptation or specialized design [28]. The concept of universal design emerged primarily with people with disability in mind. However, universal design helps everyone with support and assistance needs including the elderly, pregnant women, children, and people with a temporary illness or injury. Thus, the benefits of implementing universal design are wide [29].

In addition to "improving home safety" and "savings for government", the inclusion of universal design features in a new home brings "reducing renovation costs". A universally designed home should (1) be easy to enter; (2) be easy to move around in; (3) be capable of easy and cost-effective adaptation; and (4) be designed to anticipate and respond to the changing needs of home occupants [30]. Research suggests that incorporating universal housing design principles in advance has a minimal impact on the cost of construction. The Victorian Government of Australia has estimated the cost of including fundamental universal housing design features in a new home is 22 times cheaper than retrofitting those features into an existing home [31].

4. Toward a new comprehensive sustainable design

Homes need to be designed and constructed, so as to meet comprehensive sustainability objectives. Providing a refuge from the climate and weather, homes are expected to protect the occupants' health, safety, and well-being, as well as to reduce the burden on the natural environment.

In the twenty-first century, houses are also necessary to suit two major significant changes, namely "climate change" and "aging population". In order to curb the progress of climate change, the housing industry must intensify "mitigation" measures, such as improvement in energy efficiency and the utilization of renewable energy. It also has to strengthen "adaptation" measures against adverse effects caused by climate change, aiming to prevent or reduce damage from extreme weather events, such as intensified heat waves, rainfall, and cyclones. On the other hand, housing needs to be transformed into those which help to reduce medical and nursing care expenditure in aging population. For example, homes with higher thermal performance contribute to reducing elderly people's diseases, such as circulatory disorders in winter and heatstroke in summer. Meanwhile, incorporating accessible and universal design into houses brings various benefits, including the prevention of injuries, increase in occupants' mobility, and reduction in future renovation costs.

Houses are used for a very long time. Accordingly, in addition to usual sustainability objectives, the above emerging requirements should also be met in housing design from the beginning. Furthermore, in order to satisfy these requirements inclusively, efficient methods for new comprehensive sustainable housing design need to be developed and moreover disseminated in the society.

Author details

Kazutoshi Fujihira

Address all correspondence to: fujihira@kankyogaku.com

Institute of Environmentology, Tokyo, Japan

References

[1] Klepeis N et al. The National Human Activity Pattern Survey (NHAPS): A resource for assessing exposure to environmental pollutants. Journal of Exposure Analysis and Environmental Epidemiology. 2001;**11**:231-252. DOI: 10.1038/sj.jea.7500165

[2] Lucon O et al. Buildings. In: Edenhofer O et al., editors. Climate Change 2014: Mitigation of Climate Change. Contribution of Working Group III to the Fifth Assessment Report of the Intergovernmental Panel on Climate Change. Cambridge, United Kingdom and New York, NY, USA: Cambridge University Press; 2014. p. 671-738

[3] IPCC. Climate Change 2014: Synthesis Report. Contribution of Working Groups I, II and III to the Fifth Assessment Report of the Intergovernmental Panel on Climate Change. IPCC: Geneva, Switzerland; 2014. 151 p

[4] Sakurai H et al. Heat Stroke: Environmental Health Manual 2014 (in Japanese). Tokyo, Japan. Government of Japan: Ministry of Environment; 2014. 75 p

[5] Fujibe F. Long-term variations in heat mortality and summer temperature in Japan (in Japanese). Tenki (Weather). 2013;**60**(5):371-381

[6] Mei W, Xie S. Intensification of landfalling typhoons over the northwest Pacific since the late 1970s. Nature Geoscience. 2016;**9**:753-757. DOI: 10.1038/ngeo2792

[7] Kito A. Abnormal Weather and Global Warming (in Japanese). Iwanami Shoten: Tokyo, Japan; 2015. 208 p

[8] IPCC. Climate Change 2007: Synthesis Report. Contribution of Working Groups I, II and III to the Fourth Assessment Report of the Intergovernmental Panel on Climate Change. IPCC: Geneva, Switzerland; 2007 104 p

[9] Denton F, Wilbanks T, Abeysinghe A, Burton I, Gao Q, Lemos M, Masui T, O'Brien K, Warner K. Climate-resilient pathways: Adaptation, mitigation, and sustainable development. In: Field C et al., editors. Climate Change 2014: Impacts, Adaptation, and Vulnerability. Part

A: Global and Sectoral Aspects. Contribution of Working Group II to the Fifth Assessment Report of the Intergovernmental Panel on Climate Change. Cambridge, United Kingdom and New York, NY, USA: Cambridge University Press; 2014. p. 1101-1131

[10] Snow M, Prasad D. Climate Change Adaptation for Building Designers: An Introduction [Internet]. 2011. Available from: http://environmentdesignguide.com.au/media/misc%20 notes/EDG_66_MSa.pdf [Accessed: Mar 7, 2017]

[11] Australian Government, Department of climate change and energy efficiency. Your Home: Australia's Guide to Environmentally Sustainable Homes. Adapting to Climate Change [Internet]. 2013. Available from: http://www.yourhome.gov.au/housing/adapting-climate-change [Accessed: Mar 17, 2017]

[12] Wallace P. Agequake: Riding the Demographic Rollercoaster Shaking Business, Finance and Our World. Naperville, USA: Nicholas Brealey Publishing; 2001. 266 p

[13] United Nations Department of Economic and Social Affairs Population Division. Population Ageing and Sustainable Development [Internet]. 2014. Available from: http://www.un.org/en/development/desa/population/publications/pdf/popfacts/PopFacts_2014-4.pdf [Accessed: Mar 8, 2017]

[14] European Union. Working Group on Ageing Populations and Sustainability [Internet]. Available from: http://europa.eu/epc/working-group-ageing-populations-and-sustainability_en [Accessed: Mar 8, 2017]

[15] Cabinet Office, Government of Japan. Chapter 1: Circumstances of ageing population (1) (in Japanese) [Internet]. In: Annual Report on the Aging Society 2016 (Summary). 2016. Available from: http://www8.cao.go.jp/kourei/whitepaper/w-2016/html/gaiyou/s1_1.html [Accessed: Mar 8, 2017]

[16] Oshio T, Oguro K. Fiscal sustainability under an aging population in Japan: A financial market perspective. Public Policy Review. 2013;**9**(4):735-750

[17] Harding E. Sustainable Planning for Housing in an Ageing Population: A Guide for Regional-level Strategies. London, United Kingdom: International Longevity Centre UK; 2008. 29 p

[18] NHK Broadcasting Culture Research Institute. NHK Data Book 2015 National Time use Survey (in Japanese) [Internet]. 2016. Available from: http://www.nhk.or.jp/bunken/research/yoron/pdf/20160217_1.pdf [Accessed: Mar 9, 2017]

[19] Blackman, T. Health Risks and Health Inequalities in Housing: An Assessment Tool, Technical Report [Internet]. 2005. Available from: https://www.researchgate.net/publication/40855825_Health_risks_and_health_inequalities_in_housing_an_assessment_tool [Accessed: Mar 10, 2017]

[20] Japan Sustainable Building Consortium. Summary of CASBEE Health Checklist (in Japanese) [Internet]. 2011. Available from: http://www.ibec.or.jp/CASBEE/casbee_health/files/pamphlet.pdf [Accessed: Mar 10, 2017]

[21] Consumer Affairs Agency, Government of Japan. News Release on Bath-Related Deaths Frequently Occurring in Winter (in Japanese) [Internet]. 2016. Available from: http://www.caa.go.jp/policies/policy/consumer_safety/release/pdf/160120kouhyou_2.pdf [Accessed: Mar 10, 2017]

[22] Japan Traffic Safety Association. Number of Traffic Deaths in 2014 (in Japanese) [Internet]. 2017. Available from: http://www.jtsa.or.jp/topics/T-254.html [Accessed: Mar 10, 2017]

[23] Takahashi R. Accidents of the elderly during bathing (in Japanese). Koshu-Eisei (The Journal of Public Health Practice). 2011;**75**(8):595-599. DOI: http://dx.doi.org/10.11477/mf.1401102179

[24] Ikaga T. Surveys on Countermeasures Against Heat Stroke in Homes (in Japanese) [Internet]. 2012. Available from: http://www.mhlw.go.jp/stf/shingi/2r9852000002f13d-att/2r9852000002f1t6.pdf [Accessed: Mar 12, 2017]

[25] Ozanne-Smith J, Guy J, Kelly M, Clapperton A. The Relationship between Slips, Trips and Falls and the Design and Construction of Buildings [Internet]. 2008. Available from: http://www.monash.edu/__data/assets/pdf_file/0019/217342/muarc281.pdf [Accessed: Mar 11, 2017]

[26] National Consumer Affairs Center of Japan. Accidents in the Home Analyzed from Data of the Medical-Institution Network Project [Elderly] (in Japanese) [Internet]. 2013. Available from: http://www.kokusen.go.jp/pdf/n-20130328_3.pdf [Accessed: Mar 1, 2017]

[27] Smith S, Rayer S, Smith E. Aging and disability: Implications for the housing industry and housing policy in the United States. Journal of the American Planning Association. 2008;**74**(3):289-306. DOI: 10.1080/01944360802197132

[28] United Nations. Convention on the Rights of Persons with Disabilities [Internet]. Available from: http://www.un.org/disabilities/documents/convention/convoptprot-e.pdf [Accessed: Mar 13, 2017]

[29] Australian Agency for International Development. Accessibility Design Guide: Universal Design Principles for Australia's Aid Program. A Companion Volume to Development for All: Towards a Disability-Inclusive Australian Aid Program 2009-2014 [Internet]. 2013. Available from: http://dfat.gov.au/about-us/publications/Documents/accessibility-design-guide.pdf [Accessed: Mar 13, 2017]

[30] Australian Government Department of Social Services. National Dialogue on Universal Housing Design – Strategic Plan [Internet]. 2011. Available from: https://www.dss.gov.au/our-responsibilities/disability-and-carers/program-services/government-international/national-disability-strategy-initiatives/livable-housing-design/national-dialogue-on-universal-housing-design-strategic-plan?HTML[Accessed: Mar 13, 2017]

[31] Victorian Government of Australia, Department of Planning and Community Development. Visitable and Adaptable Features in Housing: Regulatory Impact Statement. Victorian Government of Australia: Department of Planning and Community Development; 2009. 190 p

Chapter 2

Background: Existing Japanese Systems Related to Sustainable Housing

Kazutoshi Fujihira

Additional information is available at the end of the chapter

http://dx.doi.org/10.5772/intechopen.71326

Abstract

In order to expand sustainable homes into the housing market, effective methods or systems for promoting sustainable housing design need to be widely disseminated in the society. In Japan, there are three important public systems related to sustainable housing design, namely the Housing Performance Indication System (HPIS), long-life quality housing (LQH) certification, and Comprehensive Assessment System for Built Environment Efficiency (CASBEE) for detached houses. The HPIS has stipulated the housing performance indication standards over 10 categories. The LQH system certifies houses that meet the criteria of long-life quality housing. CASBEE for detached houses comprehensively assesses and rates the sustainability of detached houses. However, there still remains room for further improvement in the set of these existing Japanese systems. First, the application of the systems to existing homes has been extremely limited until now. Second, CASBEE for detached houses, the one and only national comprehensive system, has not been used so often thus far. Moreover, readiness of the systems for emergent challenges, namely climate change and aging population, has also been insufficient.

Keywords: public system, Housing Performance Indication System, long-life quality housing, CASBEE

1. Introduction

In order to steadily increase the ratio of sustainable homes in the housing market, efficient systems or methods for promoting sustainable housing design need to be widely used in the society. In the world, a great number of systems related to sustainable housing have been developed and are used in many nations and regions; types of such systems include standards, guidelines, and assessment and rating systems [1]. Narrowing them down to major

Japanese public systems, this chapter first describes their main points. After that, it examines remaining issues in the set of these existing Japanese public systems.

2. Major existing Japanese systems related to sustainable housing

In Japan, there are three important public systems for improving the quality of housing and reducing environmental load from housing: (1) Housing Performance Indication System, (2) long-life quality housing certification, and (3) Comprehensive Assessment System for Built Environment Efficiency (CASBEE) for detached houses (**Table 1**). This section summarizes the main points of these public systems, including the aim of introduction and market penetration.

2.1. Housing Performance Indication System

2.1.1. Outline

In 2000, the Japanese government put the Housing Quality Assurance Act into operation. Before the enforcement of this law, there were no common rules to indicate housing performance. Accordingly, it was difficult for consumers to compare the quality of houses in the housing market. On the other hand, there was little incentive for housing producers to compete for higher performance [2]. In order to solve these problems and consistently assure the quality of houses, the Housing Quality Assurance Act has prescribed the Housing Performance Indication System (HPIS) or *Jutaku Seino Hyoji Seido* [2].

2.1.2. Indication items

The HPIS has stipulated the "Japan Housing Performance Indication Standards (JHPIS)" or *Nihon Jutaku Seino Hyoji Kijun* and the indication procedure standards [2]. Based on these standards, registered third party organizations assess housing performance and indicate its results.

The JHPIS has specified items to be indicated for homes over 10 categories [2]. The 10 categories and main items are as follows:

1. Structural stability (Resistance to earthquakes, etc.),
2. Fire safety (Fire resistance),
3. Reduction of deterioration (Measures against deterioration),
4. Consideration for maintenance (Measures for maintenance),
5. Thermal environment/energy (Thermal insulation performance or primary energy consumption),
6. Indoor air environment (Measures against formaldehyde),

7. Luminous and visual environment (Ratio of window area to floor area),
8. Acoustic environment (Sound insulation performance, etc.),
9. Consideration for the aged and others (Accessibility),
10. Security (Measures against intrusion at openings).

System name	Housing Performance Indication System	Long-life quality housing certification	CASBEE for detached houses
Essence	• Stipulation of housing performance indication standards • Assessment & indication of housing performance by registered third-party organizations	• Certification of long-life quality houses	• Comprehensive assessment of built environment efficiency • Rating and certification based on the assessment
Aim of introduction	• Enabling consumers to compare performance of houses • Builders' competition for higher performance • Assurance of housing quality	• Prolongation of housing life span • Improvement of quality of housing • Reduction of people's economic burden and environmental burden	• Promoting sustainable housing design • Measures against global environmental problems
Housing type	• New (2000-)/Existing (2002-) • Detached/Collective	• New (2009-)/Existing (2016-) • Detached/Collective	• New (2007-)/Existing (2011-) • Detached
Categories or items of indication/assessment for detached houses	• Structural stability • Fire safety • Reduction of deterioration • Consideration for maintenance • Thermal environment/energy • Indoor air environment • Luminous and visual environment • Acoustic environment • Consideration for the aged and others • Security	• Measures against deterioration • Resistance to earthquakes • Measures for maintenance • Indoor temperature and energy saving • Consideration for landscape • Total floor area • Planning for future checking & maintenance	• Comfortable, healthy & safe indoor environment • Durability for long-term use • Consideration for landscape and ecosystem • Energy and water saving • Conservation of resources and reduction of waste • Consideration for the global, local and surrounding environment
Incentives for system users		• Tax reduction (Income tax, real estate tax, etc.) • Support in taking housing mortgages	
Market penetration: Accumulated number of houses with indication/certification	[New detached houses]: 927,216 (2000–2014) [Exist. detached houses]: 1004 (2002–2012)	[New detached houses]: 576,068 (2009–2014)	[New detached houses]: 120 (2008–2016)

Table 1. Main points of the Housing Performance Indication System, long-life quality housing certification, and CASBEE for detached houses.

In 2015, the classification of the items into mandatory and optional items was revised. As a result, the number of the optional items has sharply increased. At present, all the items in the six categories, fire safety, indoor air environment, luminous and visual environment, acoustic environment, consideration for the aged and others, and security, are optional [3].

2.1.3. Market penetration

At the beginning of the adoption of the HPIS, the objects of assessment were limited to only new homes. In 2002, the HPIS was amended, so as to extend its coverage to existing homes [2].

After the start of the HPIS in 2000, the annual penetration rate into the new housing market steadily increased and reached 20% in 2006. After that, the annual penetration rate fluctuated between 19 and 24% [4]. According to the newest statistical data, in 2014, the total number of officially assessed new homes amounted to 196,021, which is equivalent to 22.3% of the total housing starts. The breakdown of these is as follows: (1) detached houses: 88,312 houses, 21.6% and (2) collective houses: 107,709 houses, 22.9% [4].

Meanwhile, this system has rarely been utilized on existing houses. As of the end of fiscal 2012, the cumulative total number of the officially assessed existing homes has been only 3433. The breakdown of this is 1004 detached houses and 2429 collective houses [5].

2.2. Long-life quality housing certification

2.2.1. Outline

In Japan, the life span of houses has been short, as compared to those in Europe and the United States. The average elapsed period of Japanese houses from the construction to the demolition is only about 30 years. Frequent construction and demolition increase materials and energy consumption for construction as well as waste. Moreover, shorter life span of houses increases people's housing expenses [6]. In order to reduce such burden on the natural environment and economic burden on people as well as improve the quality of housing, in 2009, the Japanese government introduced a new housing certification system named "Long-life quality housing (LQH)," in Japanese *Choki Yuryo Jutaku* [6, 7].

For quite a while after the start of the LQH certification system, it was applied to only new homes. In 2016, the government extended its coverage to existing homes [7].

2.2.2. Certification criteria

The Ministry of Land, Infrastructure, Transport and Tourism has stipulated the criteria for the LQH certification [8]. In the case of "detached houses," the LQH certification criteria and their essence are as follows:

1. Measures against deterioration (measures for at-least-100-year continuous use of framework),
2. Resistance to earthquakes (structure with 1.25 times the strength stipulated in the building code or base-isolated structure),

3. Readiness for maintenance and replacement (measures for easy maintenance and replacement of piping and interior),
4. Indoor temperature and energy saving (sufficient thermal insulation performance),
5. Local environment (consideration for landscape),
6. Total floor area (75 m^2 or more), and
7. Maintenance plan (planning for future checking and maintenance) [8].

In the case of "collective houses," the LQH certification requires two more criteria, in addition to the above seven criteria. The two criteria are "preparedness for future change of room arrangement" and "sufficient space in common areas for improving accessibility" [8].

2.2.3. Market penetration

The LQH certification system considerably penetrates into the housing market, especially new detached houses. In 2009, the first year of the system introduction, the penetration rate into new detached houses reached 17.2% immediately. In 2010, it increased to 23.8%. After that the annual penetration rate into new detached houses fluctuates between 23 and 24%. As of the end of fiscal 2014, the total number of the certified new detached houses amounted to 576,068. On the other hand, the penetration into the collective housing market is in a sluggish state; as of the end of fiscal 2014, the total number of the certified collective housing units is only 15,939 [9].

A significant reason of rapid dissemination of LQH certification is preferential treatment to the system users. The construction costs of the LQH-certified houses are considerably expensive compared with usual houses. Therefore, aiming to promote the LQH certification, the government has granted various kinds of preferential treatment to people who acquire the LQH-certified houses [10]. Such treatment includes reduction of income tax, real estate tax, real estate acquisition tax, and registration and license tax, as well as support in taking housing mortgages through the Government Housing Loan Corporation [10]. Moreover, there is also preferential treatment to small and medium-sized homebuilders that provide the LQH-certified homes to be concrete subsidies from the government [11].

2.3. CASBEE for detached houses

2.3.1. Outline

From around 1990, various methods for promoting green or sustainable building design appeared in the world, reflecting growing awareness of seriousness of global environmental problems [12, 13]. In 1990, the BREEAM (Building Research Establishment Environmental Assessment Method) was developed in the United Kingdom, as the world's first sustainability assessment method for buildings [14]. In the U.S., the U.S. Green Building Council was founded in 1993, with a mission to promote sustainability-focused practices in the building and construction industry [15]. In 2000, the U.S. Green Building Council unveiled the green building rating system named LEED (Leadership in Energy and Environmental Design) [15]. In the middle of this global trend, Comprehensive Assessment System for Built Environment Efficiency (CASBEE) appeared in Japan [12, 13].

CASBEE is a method for comprehensively assessing and rating the sustainability of buildings and built environment. In 2001, CASBEE was developed in committee set up in the Institute for Building Environment and Energy Conservation (IBEC) under the initiative of the Ministry of Land, Infrastructure Transport and Tourism [16]. Recently, the Japan GreenBuild Council and the Japan Sustainable Building Consortium have been continuously developing and updating the CASBEE systems. The CASBEE systems have been named according to the specific purposes, such as "for Building," "for Urban Development," "for Cities," and "for Detached Houses." In addition, "for Detached Houses" includes two versions: "CASBEE for Detached Houses (New Construction)," which was developed in 2007, and "CASBEE for Detached Houses (Existing Building)," which was added in 2011, as a version for existing detached houses [13, 17].

2.3.2. *Assessment categories and items*

CASBEE covers four assessment fields: (1) Energy efficiency, (2) Resource efficiency, (3) Local environment, and (4) Indoor environment, which are largely the same as the assessment fields of other major comprehensive methods such as BREEAM and LEED [13, 18]. However, in order to assess environmental performance more precisely, the above four assessment fields have been uniquely recategorized into Built Environment Quality (Q) and Built Environment Load (L) (**Figure 1**). "Q" is further divided into three subcategories: [Q1] Indoor environment, [Q2] Quality of service, and [Q3] Outdoor environment on site. Similarly, "L" is divided into [L1] Energy, [L2] Resources and materials, and [L3] Off-site environment. The quotient of "Q" divided by "L" represents building environmental performance with a single score of Built Environment Efficiency (BEE). According to the BEE score, the buildings are classified into five grades: S (Excellent), A (Very good), B+ (Good), B− (Fairly poor), and C (Poor) [13, 18].

In case of "CASBEE for Detached Houses," 21 middle-level assessment items and 54 scoring items are allotted over the total six subcategories [13]. The following shows the middle-level assessment items by each subcategory:

[Q1] Comfortable, healthy, and safe indoor environment: "Heat and cold," "Health, safety, and security," "Brightness," and "Quietness,"

[Q2] Durability for long-term use: "Basic performance for the duration of long-term use," "Maintenance and management," and "Service ability,"

[Q3] Consideration for the townscape and ecosystem: "Consideration for townscape and view," "Creation of biological environment," "Local safety and security," and "Utilization of local resources and preservation of the architectural/dwelling cultural heritage,"

[L1] Energy and water conservation: "Energy conservation by means of improvements to the house," "Energy conservation by means of improvements to facilities," "Water conservation," and "Improvements to the maintenance, management and operation system,"

[L2] Conservation of resources and reduction of waste: "Use of resource-saving materials and less waste-producing materials," "Waste reduction at the production/construction stages," and "Recyclability," and,

[L3] Consideration for the global, local and surrounding environment: "Consideration for the global environment," "Consideration for the local environment," and "Consideration for the surrounding environment" [13].

2.3.3. Market penetration

According to the organizer of CASBEE, CASBEE is broadly used by many construction companies, design offices, real-estate developers, etc. as a voluntary-basis evaluation tool for checking the environmental performance of their buildings [19]. However, statistical and survey data indicate that CASBEE is not so widespread, especially among smaller companies.

The number of the CASBEE certification is extremely limited, as compared with other major assessment and certification systems, such as BREEAM and LEED. As of March 2017, the total number of the CASBEE-certified buildings/projects is approximately 600, which includes 120 of the "CASBEE for Detached Houses" certified homes. All of the CASBEE-certified homes are new homes; the number of the certified existing homes is still zero [19, 20].

On the other hand, the number of BREEAM and LEED certification is larger than that of CASBEE certification by orders of magnitude. According to the home page of BREEAM, globally there are more than 559,900 BREEAM-certified developments, and almost 2,262,100 buildings registered for assessment, since it was launched [14]. Similarly, "USGBC Statistics" shows that there are more than 36,400 certified commercial projects and more than 121,400 certified LEED for Homes residential units, as of January 2017 [21].

Meanwhile, a questionnaire to home builders suggests that CASBEE for detached houses is not used so often, or not even known so much, especially by smaller organizations. According to the analysis based on 252 valid survey responses from Japanese housing construction companies, about half of the respondents answered that they were unaware of CASBEE [22]. More specifically, "51.6%" of the respondents reported that they "had never heard of" CASBEE for detached houses, while "34.1%" of respondents answered that they "had heard of it but had never used" it, and "14.3%" of respondents reported that they "had used" it. In addition, the

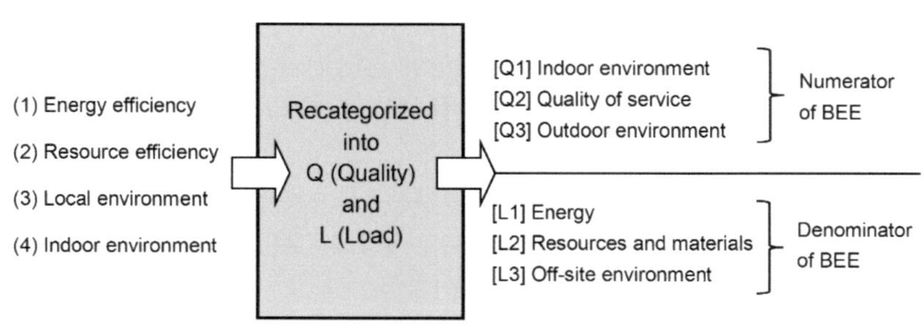

Figure 1. The core concept of CASBEE [13].

ratio of answering "Never heard of" was higher among the respondents from smaller sized housing companies; the ratio of reporting "Used" was higher among the respondents from larger sized companies [22].

3. Remaining issues in the existing Japanese systems

As described in the last section, the three public systems, namely the HPIS, LQH certification, and CASBEE for detached houses, have been developed and are used in Japan. However, there remain issues in the dissemination of these systems as well as readiness for emergent challenges, such as climate change and aging population. This section examines these remaining issues from three perspectives: (1) application to new and existing homes, (2) issue of CASBEE dissemination, and (3) readiness for emergent challenges.

3.1. Application to new and existing homes

As shown in the market penetration data, the HPIS and LQH certification have considerably penetrated into the new housing market. The penetration rates into new detached houses rapidly increased and reached 20-plus percent. However, after that the penetration rates seem to have hit a ceiling at this level. From now on, it is necessary to provide momentum for further rises in the penetration rates.

Meanwhile, the applications of the systems to existing homes have been extremely limited thus far. The HPIS has rarely been utilized on existing houses, and the LQH certification of existing houses has just started. It is obvious that these systems need to be disseminated into massive amounts of the existing housing stock.

3.2. Issue of CASBEE dissemination

CASBEE for detached houses is the only national comprehensive method for sustainable housing design. However, it has not been used so often, and the number of the certification is extremely limited. There appears to be two main reasons why CASBEE does not become widespread, that is to say, "complexity" and "shortage of incentive to obtain certification."

CASBEE is described as elaborate and hard to understand, whereas LEED is reputed as simple and easy to understand [23]. A significant cause of CASBEE's complexity is its unique core concept. As already shown in **Figure 1**, the original four categories have been recategorized into "Q" and "L." This recategorization has inevitably made the system complex and difficult to understand.

The second reason is shortage of incentive to obtain CASBEE certification, as compared with LEED. In the U.S., many states and local governments have adopted various types of direct incentives for "green buildings," including LEED-certified buildings. Examples of such incentives are tax credits, revolving loan funds, and expedited review/permitting processes [24, 25]. In addition, in the LEED system, only buildings with considerably high environmental performance can be certified (**Figure 2**, left); therefore, LEED-certified buildings are naturally recognized as green buildings [23]. Furthermore, in the areas where LEED is widespread, there has been building a consensus that green buildings have competitive superiority in the real estate market. In other words, people in such areas have been sharing information on studies which show that certified green buildings outperform over traditional buildings in asset value, rent, and occupancy rate [26–28]. In this way, there are also social incentives to obtain a LEED certification, in addition to the direct incentives.

On the other hand, there is little incentive to obtain CASBEE certification. Originally, CASBEE systems certify all the assessed buildings and rate them into five ranks: S, A, B+, B-, and C. Accordingly, as shown in the right of **Figure 2**, CASBEE-certified buildings do not necessarily mean green buildings. In addition, governments hardly offer direct incentives, such as tax credits, to developers or owners of grade "S" or "A" certified buildings. Moreover, there has also been little shared understanding that green buildings have competitive advantage in the real estate market of Japan thus far [29].

3.3. Readiness for emergent challenges

As "climate change" becomes aggravated, houses need to take further mitigation measures as well as adaptation measures. The LQH certification requires higher thermal insulation performance of the housing envelope, which is significant as both mitigation and adaptation measures. However, it refers to neither equipment for harnessing renewable energy nor energy efficiency of apparatus such as water heaters. CASBEE for detached houses refers to all of these items, but there is a problem which the system itself has not been

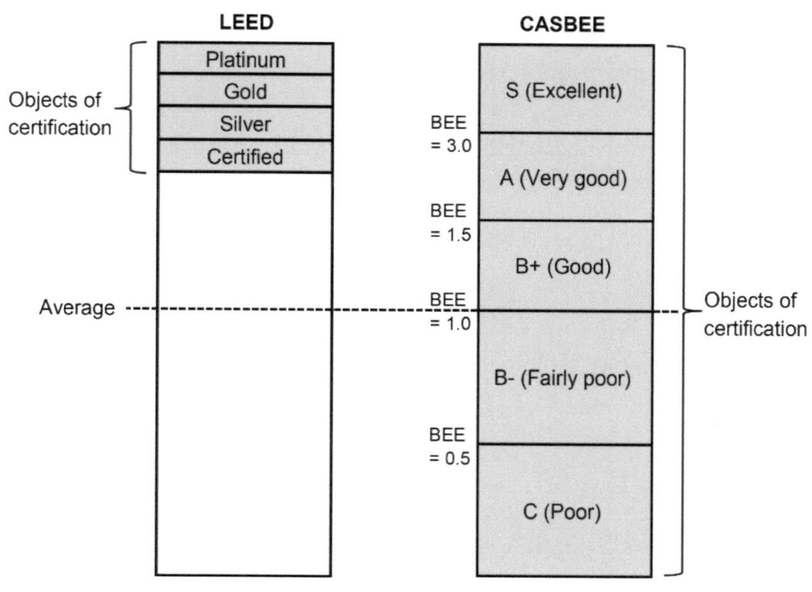

Figure 2. Rating and certification of LEED and CASBEE.

used so often until now. Meanwhile, none of the three existing systems have included any descriptions of strengthening adaptation measures thus far.

On the other hand, readiness for "aging population" in the housing systems has been sluggish, although Japan is the fastest aging country in the world. The LQH certification requires detached houses to take neither accessibility nor universal design. The HPIS includes "Consideration for aged and others," which refers to accessibility, as one of the categories to be indicated. However, in 2015, "Consideration for aged and others" changed from a mandatory category to be indicated to an optional one. Such a system amendment even appears to be against the time.

4. Necessity of more efficient systems for promoting sustainable design

From around 2000, public systems for improving housing performance were developed and have been used in Japan. Especially the three systems, namely the HPIS, LQH certification, and CASBEE detached houses, occupy an important position in the Japanese housing policy. The HPIS has stipulated standards for assessing housing performance. The LQH system certifies houses which meet the criteria of long-life quality housing. CASBEE for detached houses comprehensively assesses and rates the sustainability of detached houses.

However, there still remain issues in the set of these existing systems. First, the application of the systems to existing homes has been extremely limited until now. Second, CASBEE for detached houses, the one and only national comprehensive system, has not penetrated much into the housing industry, especially among smaller companies, thus far. Moreover, readiness of the systems for the emergent challenges is insufficient. In particular, taking up accessibility and universal design into detached houses has been sluggish, although Japan is the world's fastest progressing aging country. It is obvious that the set of the existing Japanese public housing systems need to be improved toward more efficient one.

Author details

Kazutoshi Fujihira

Address all correspondence to: fujihira@kankyogaku.com

Institute of Environmentology, Tokyo, Japan

References

[1] Carmody J, Weber W, Jacobson R. Center for Sustainable Building Research (CSBR) University of Minnesota. Design Guidelines for Sustainable Housing [Internet]. 2009. Available from: http://www.csbr.umn.edu/download/KoreaHousingProjectCombinedFinalReport.pdf [Accessed: May 17, 2017]

[2] Ministry of Land, Infrastructure, Transport and Tourism. Summary of the Housing Quality Assurance Act (in Japanese) [Internet]. 2007. Available from: http://www.mlit.go.jp/common/000052921.pdf [Accessed: Mar 27, 2017]

[3] Ministry of Land, Infrastructure, Transport and Tourism. Housing Performance Indication System: Revision of the Classification of the Indication Items into Mandatory and Optional Items (in Japanese) [Internet]. 2015. Available from: http://www.mlit.go.jp/common/001090813.pdf [Accessed: Mar 27, 2017]

[4] Housing Performance Assessment and Indication Association (Jutaku Seino Hyoka Hyoji Kyokai). Chronological Change of the Penetration Rate of the Housing Performance Indication System (in Japanese) [Internet]. 2015. Available from: https://www.hyouka-kyoukai.or.jp/kokai/data/fukyuritsu.pdf [Accessed: Mar 27, 2017]

[5] Ministry of Land, Infrastructure, Transport and Tourism. Usage Situation of the Housing Performance Indication System (in Japanese) [Internet]. Available from: http://www.mlit.go.jp/common/001047038.pdf [Accessed: Mar 27, 2017]

[6] Ministry of Land, Infrastructure, Transport and Tourism. Summary of the Basic Act for Housing (in Japanese) [Internet]. Available from: http://www.mlit.go.jp/jutakukentiku/house/singi/syakaishihon/bunkakai/14bunkakai/14bunka_sankou04.pdf [Accessed: Mar 27, 2017]

[7] Ministry of Land, Infrastructure, Transport and Tourism. Information on the Law for the Promotion of Long-Term Quality Housing (in Japanese) [Internet]. 2008. Available from: http://www.mlit.go.jp/jutakukentiku/house/jutakukentiku_house_tk4_000006.html [Accessed: Mar 27, 2017]

[8] Ministry of Land, Infrastructure, Transport and Tourism. Summary of Long-Term Quality Housing Certification Standards (in Japanese) [Internet]. Available from: http://www.mlit.go.jp/common/000041415.pdf [Accessed: Mar 27, 2017]

[9] Ministry of Land, Infrastructure, Transport and Tourism. State of the Certification of Long-Term Quality Construction Plans Based on the Law for the Promotion of Long-Term Quality Housing (in Japanese) [Internet]. 2015. Available from: http://www.mlit.go.jp/report/press/house04_hh_000589.html [Accessed: Mar 1, 2017]

[10] Ministry of Land, Infrastructure, Transport and Tourism. Summary of the Law for the Promotion of Long-Term Quality Housing, Tax System, and Mortgage (in Japanese) [Internet]. Available from: http://www.mlit.go.jp/common/001158174.pdf [Accessed: Mar 27, 2017]

[11] Region-based projects for greening housing: section for supporting the long-life type, etc. Long-life-type long-term quality housing (in Japanese) [Internet]. 2015. Available from: http://www.chiiki-grn-chojyu.jp/choujumyou28/manual/index.html [Accessed: Mar 27, 2017]

[12] Japan GreenBuild Council, Japan Sustainable Building Consortium. CASBEE: The Background to CASBEE Development [Internet]. Available from: http://www.ibec.or.jp/CASBEE/english/backgroundE.htm [Accessed: Mar 28, 2017]

[13] Murakami S, Iwamura K, Cole R. CASBEE: A Decade of Development and Application of an Environmental Assessment System for the Built Environment. Tokyo, Japan: Institute for Building Environment and Energy Conservation; 2014 297 p

[14] Building Research Establishment. BREEAM: Why BREEAM? [Internet]. 2017. Available from: http://www.breeam.com/why-breeam [Accessed: Mar 28, 2017]

[15] U.S. Green Building Council. About: Our History [Internet]. Available from: http://www.usgbc.org/About [Accessed: Mar 28, 2017]

[16] Japan Sustainable Building Consortium, Japan GreenBuild Council. CASBEE for Home (Detached House) Technical Manual. 2007th ed. Tokyo, Japan: Institute for Building Environment and Energy Conservation; 2008 263 p

[17] Japan GreenBuild Council, Japan Sustainable Building Consortium. CASBEE: An Overview of CASBEE [Internet]. Available from: http://www.ibec.or.jp/CASBEE/english/overviewE.htm [Accessed: Mar 28, 2017]

[18] Japan GreenBuild Council, Japan Sustainable Building Consortium. CASBEE: The Assessment Method Employed by CASBEE [Internet]. Available from: http://www.ibec.or.jp/CASBEE/english/methodE.htm [Accessed: Mar 28, 2017]

[19] Japan GreenBuild Council, Japan Sustainable Building Consortium. CASBEE: Dissemination of CASBEE in Japan [Internet]. Available from: http://www.ibec.or.jp/CASBEE/english/statistics.htm [Accessed: Mar 28, 2017]

[20] Japan GreenBuild Council, Japan Sustainable Building Consortium. CASBEE: List of Certified Detached Houses [Internet]. Available from: http://www.ibec.or.jp/CASBEE/DH_certification/certified_homes_list.htm [Accessed: Mar 28, 2017]

[21] U.S. Green Building Council. USGBC Statistics [Internet]. 2017. Available from: http://www.usgbc.org/articles/usgbc-statistics [Accessed: Mar 29, 2017]

[22] Sasatani D, Bowers T, Ganguly I, Eastin I. Adoption of CASBEE by Japanese house builders. Journal of Green Building. 2015;**10**(1):186-201. DOI: http://dx.doi.org/10.3992/jgb.10.1.186

[23] Hiramatsu H. Trend of global-based building standards (LEED etc.) and differences between them and CASBEE (in Japanese). Journal of Real Estate Securitization. 2011;**3**:44-50

[24] American Institute of Architects. Local Leaders in Sustainability: Green Incentives – State and Local Green Building Incentives [Internet]. Available from: http://www3.cec.org/islandora-gb/en/islandora/object/greenbuilding%3A144 [Accessed: Mar 31, 2017]

[25] U.S. Green Building Council. Good to Know: Green Building Incentive Strategies [Internet]. 2014. Available from: http://www.usgbc.org/articles/good-know-green-building-incentive-strategies-0 [Accessed: Mar 31, 2017]

[26] Institute for Market Transformation & Appraisal Institute. Green Building and Property Value: A Primer for Building Owners and Developers [Internet]. 2013. Available from: https://www.appraisalinstitute.org/assets/1/7/Green-Building-and-Property-Value.pdf [Accessed: Mar 1, 2017]

[27] Dodge Data & Analytics. World Green Building Trends 2016: Developing Markets Accelerate Global Green Growth [Internet]. Available from: http://fidic.org/sites/default/files/World%20Green%20Building%20Trends%202016%20SmartMarket%20Report%20FINAL.pdf [Accessed: Mar 1, 2017]

[28] Bentall Kennedy. Building Certifications [Internet]. 2017. Available from: http://cr.bentall-kennedy.com/Environment/BuildingCertifications [Accessed: Mar 31, 2017]

[29] Hori, M. Present Situation and Future Prospect of Building Environmental Performance Assessment (in Urban Study Vol. 51, in Japanese) [Internet]. 2010. Available from: http://www.minto.or.jp/print/urbanstudy/pdf/u51_04.pdf [Accessed: Mar 31, 2017]

Basic Schemes: Preparations for Applying Control Science to Sustainable Design

Kazutoshi Fujihira

Additional information is available at the end of the chapter

http://dx.doi.org/10.5772/intechopen.71325

Abstract

It is the ultimate goal for humankind to deal with various problems and achieve sustainability. Control science can be applied to all goal-oriented tasks and has already produced remarkable results. Accordingly, applying control science to the task of achieving sustainability is a rational and reliable approach. In order to apply control science to sustainability issues, our first study has shown the "basic control system for sustainability" as well as the "model of sustainability." After that, in order to identify system components of practical control systems for promoting sustainable design, we have devised "two-step preparatory work for sustainable design." The two steps of this preparatory work are "determining the relationships between the standard human activities and sustainability" and "sustainability checkup on human activities as an object."

Keywords: control science, control system, model of sustainability, two-step preparatory work, sustainable design

1. Introduction

In order to develop a methodology for promoting sustainable housing design, we have applied the science of control. The reason is that accomplishing sustainability is regarded as the ultimate goal-oriented challenge, and control science is suitable for all goal-oriented tasks.

In order to deal with a variety of problems and achieve sustainability, "human beings must **control** their activities appropriately" [1], I wrote in my book, *An Introduction to Environmentology*, which was published in 1999. In 2001, I conceived the idea of applying the science of "control" to this ultimate challenge of humankind. "Control" is generally defined as "purposive influence toward a predetermined goal" [2]. Control science can be utilized for all goal-oriented tasks [3]. In fact, control science has been applied to various fields, including engineering, medicine, agriculture, and economics; particularly, control engineering has a long history and has produced extraordinary results [3]. Thus, it is a logical and reliable approach to apply control science to the challenge of accomplishing sustainability.

Meanwhile, the application of control science to sustainable design has required us to develop basic schemes in advance for preparations. First, we have demonstrated the basic control system for sustainability as well as the model of sustainability, so as to facilitate the utilization of control science for various sustainability issues. Moreover, we have devised the two-step preparatory work for sustainable design. The following sections illustrate these basic schemes for applying control science to sustainable design.

2. Basic control system for sustainability

After starting research on applying control science to sustainability issues, first of all, we showed the basic control system for sustainability. Recently, we have revised it, based on the IPCC's recognition, namely the necessity of adaptation measures for sustainability [4–6].

Figure 1 demonstrates the revised version of the basic control system for sustainability. "Controlled objects" are human activities which need to be controlled [7–9]. "Disturbances" are adverse effects on controlled objects which are caused by environmental, social, or economic problems [7–9]. Concrete instances of the disturbances include floods or landslides resulting from environmental destruction, harmful influences caused by environmental pollution, and a variety of impacts resulting from climate change [7–9]. Furthermore, "adaptation" has been added as the course from "disturbances" to "sustainability," on the basis of the recent recognition of the IPCC [7]. "Controlled variables" are the variables that relate to the human activities and need to be controlled for primarily solving or preventing the problems or adapting to disturbances [7].

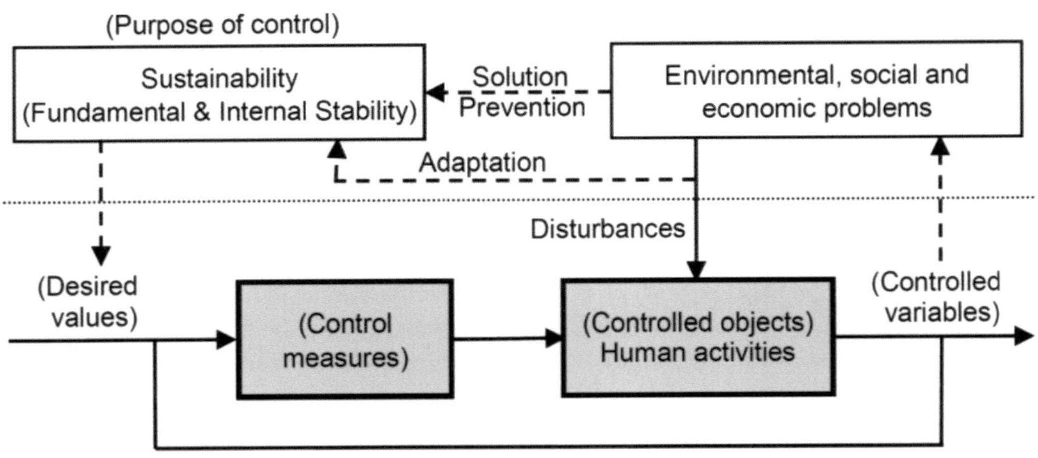

Figure 1. Basic control system for sustainability [7].

"Desired values" are derived from the purpose of control, namely sustainability [7–9]. The model of sustainability (**Figure 2**) shows that sustainability needs both "Internal Stability" and "Fundamental Stability," in order to achieve the long-term well-being of all humankind, or ultimate goal, within the finite global environment and natural resources, or absolute limitations [7–9]. "Internal Stability" means social and economic stability; the conditions for Internal Stability are health, safety, mutual help, and self-realization, which are important for well-being of humans [7–9]. On the other hand, "Fundamental Stability" means environmental stability and a stable supply of necessary goods; the conditions for Fundamental Stability are environmental preservation and the sustainable use of natural resources [7–9].

The control objective of the basic control system for sustainability is to adjust the controlled variables to their desired values [7]. Moreover, the control system requires designing and implementing "control measures" or measures for attaining the control objective [7].

3. Two-step preparatory work for sustainable design

In order to identify a control objective, system designers must identify controlled variables and their desired values. Therefore, preparatory work for designing control measures is primarily intended to identify controlled variables and their desired values. This preparatory work con-

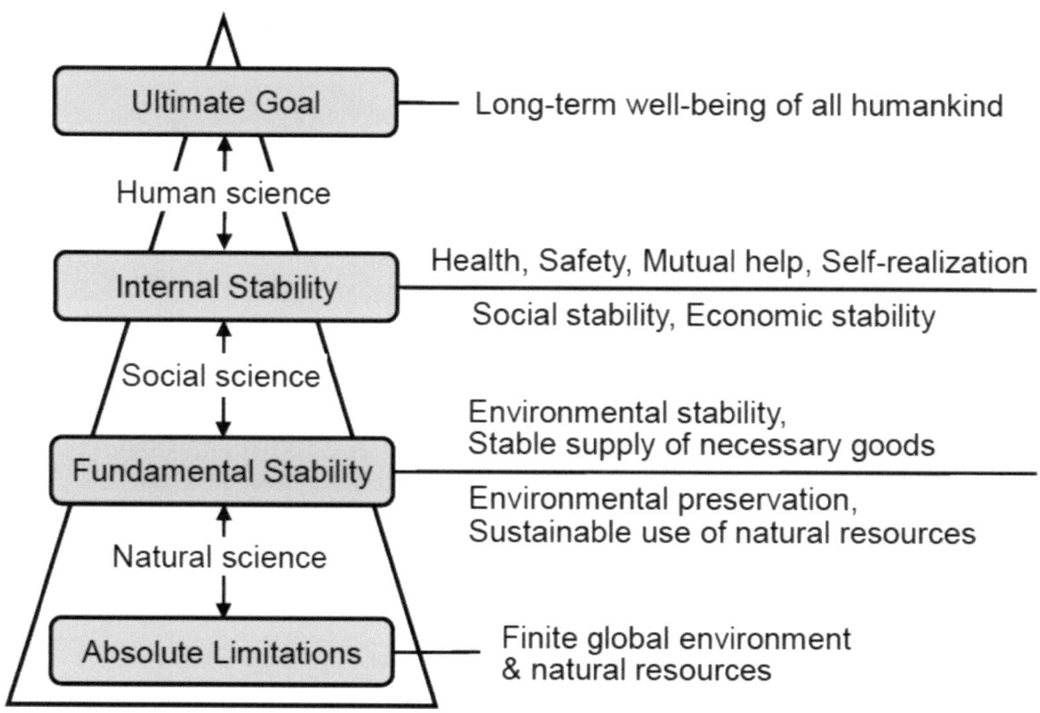

Figure 2. Model of sustainability [7].

sists of two steps: (1) determining the relationships between the standard human activities and sustainability and (2) sustainability checkup on human activities as an object [10].

3.1. Determining the relationships between the standard human activities and sustainability

The first step aims to comprehensively determine the relationships between the standard human activities and sustainability [10]. The standard human activities mean typical human activities among human activities which belong to one category. **Figure 3** demonstrates the concept of this step [10].

The first step starts with selecting important elements from the standard human activities [10]. Human activities in one category include almost the same elements. Therefore, at first, system designers select such common elements from the standard human activities [10]. In this connection, if one or more factors which influence the selection can be found, the selection process will become more efficient [10]. In addition, the elements which are considered to be closely related to sustainability should always be selected as important elements, no matter whether they are common in the present situation [10]. For instance, when "home" is

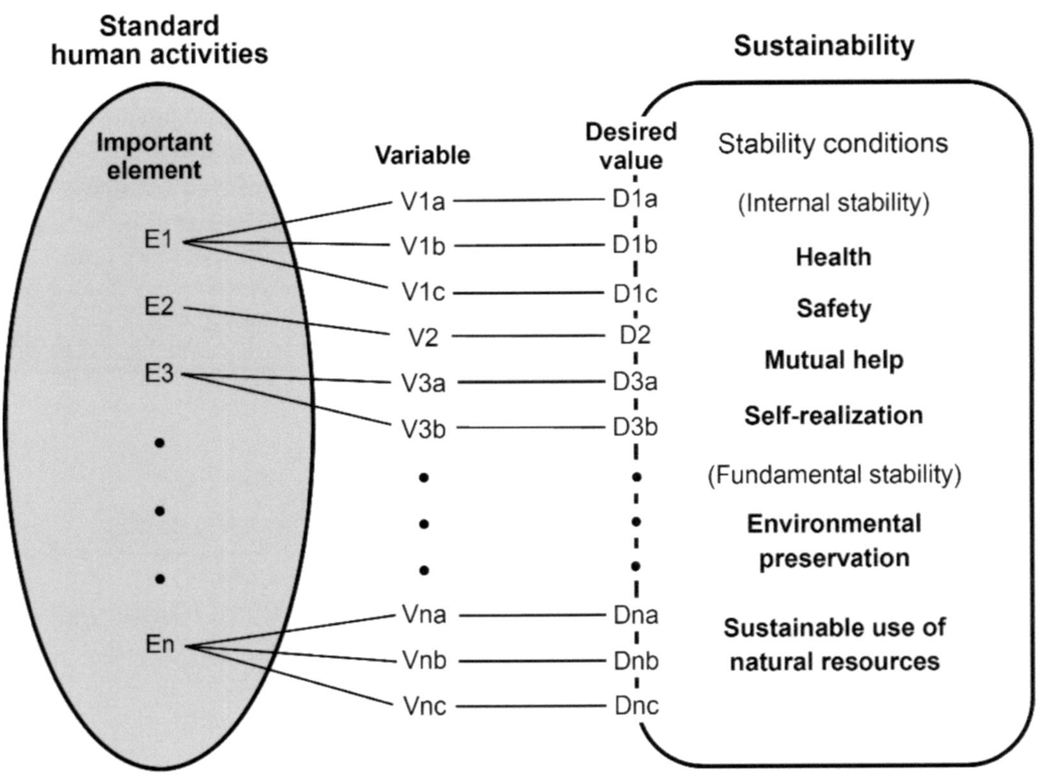

Figure 3. Concept diagram of determining the relationships between the standard human activities and sustainability [10].

chosen as a category of human activities, "equipment for rainwater use" need to be selected as an important element, even though it is not common in current ordinary homes [10].

After system designers select important elements, they determine the relationships between the selected elements and sustainability [10]. This work is composed of three processes: (1) considering the relationships between each element and the stability conditions for both Internal Stability and Fundamental Stability, including health, safety, and environmental preservation; (2) identifying variables which can indicate the degree of stability; (3) setting the variables' desired values that can meet relevant stability conditions [10]. As demonstrated in **Figure 3**, the number of variables that connect to one element is not necessarily one but can be many [10]. In addition, system designers need to identify variables and set their desired values, based on the most recent technology, scientific knowledge, and social conditions [10].

3.2. Sustainability checkup on human activities as an object

The second step is "sustainability checkup on human activities as an object" [10]. The second step starts with the measurement or estimation of the above-mentioned variables of human activities as an object [10]. Subsequently, comparing the measured or estimated values with the desired values, system designers assess the degree of stability [10].

After the comparison and assessment, the variables that fall below the desired values are necessary to be identified as "controlled variables" [10]. In addition, human activities as an object that includes controlled variables are naturally identified as a "controlled object" [10].

Author details

Kazutoshi Fujihira

Address all correspondence to: fujihira@kankyogaku.com

Institute of Environmentology, Tokyo, Japan

References

[1] Fujihira K. An Introduction to Environmentology (Kankyogaku Nyumon, in Japanese). Tokyo, Japan: Nihon Keizai Shuppansha; 1999. 345 p

[2] Beniger J. The Control Revolution—Technological and Economic Origins of the Information Society. Cambridge, Massachusetts, USA: Harvard University Press; 1986. 508 p

[3] Osuka K, Adachi S. Approach to Systems Control (in Japanese). Tokyo, Japan: Corona Publishing; 1999. 177 p

[4] IPCC. Climate Change 2007: Synthesis Report. Contribution of Working Groups I, II and III to the Fourth Assessment Report of the Intergovernmental Panel on Climate Change. Geneva, Switzerland: IPCC; 2007. 104 p

[5] IPCC. Climate Change 2014: Synthesis Report. Contribution of Working Groups I, II and III to the Fifth Assessment Report of the Intergovernmental Panel on Climate Change. Geneva, Switzerland: IPCC; 2014. 151 p

[6] Denton F, Wilbanks T, Abeysinghe A, Burton I, Gao Q, Lemos M, Masui T, O'Brien K, Warner K. Climate-resilient pathways: Adaptation, mitigation, and sustainable development. In: Field C et al., editors. Climate Change 2014: Impacts, Adaptation, and Vulnerability. Part A: Global and Sectoral Aspects. Contribution of Working Group II to the Fifth Assessment Report of the Intergovernmental Panel on Climate Change. Cambridge, United Kingdom and New York. NY, USA: Cambridge University Press; 2014. p. 1101-1131

[7] Fujihira K. System control for sustainability: Application to building design. In: Thomas C, editor. Complex Systems, Sustainability and Innovation. Rijeka, Croatia: InTech; 2016. p. 37-55. DOI: http://dx.doi.org/10.5772/65875

[8] Fujihira K, Osuka K, Yoshioka T, Hayashi N. An educational methodology for sustainable development applying control theory and confirming its validity (in Japanese). Environmental Education. 2008;**18**(1):17-28

[9] Fujihira K, Osuka K. An educational methodology for sustainable development. In: ICCAS-SICE 2009 Final Program and Papers; 18-21 August, 2009; Fukuoka, Japan. Tokyo, Japan: SICE; 2009

[10] Fujihira K. An approach to sustainable development by applying control science. In: Ghenai C, editor. Sustainable Development – Policy and Urban Development – Tourism, Life Science, Management and Environment. Rijeka, Croatia: InTech; 2012. p. 299-318

Chapter 4

Methodology of Applying Control Science to Sustainable Housing Design

Kazutoshi Fujihira

Additional information is available at the end of the chapter

http://dx.doi.org/10.5772/intechopen.71324

Abstract

The previous chapter has demonstrated the "basic control system for sustainability", "model of sustainability", and "two-step preparatory work for sustainable design". Based on these basic schemes, this chapter shows the methodology of applying control science to sustainable home design. First, using two factors, that is, "material" and "space", we identify important elements of home. Next, we provide the two-step preparatory work for sustainable home design, namely, (1) determining the relationships between the standard home and sustainability and (2) sustainability checkup on a home as an object. After that, we derive "sustainable design guidelines" from step 1 and "sustainability checklist" from step 2, respectively. Finally, we compose the "control system for promoting sustainable home design" in which the sustainable design guidelines and sustainability checklist are incorporated. This practical control system demonstrates sustainable design processes for both new and existing homes.

Keywords: control system, material elements, spatial elements, sustainable design guidelines, sustainability checklist

1. Introduction

Chapter 3 has shown the "basic control system for sustainability," "model of sustainability," and "two-step preparatory work for sustainable design." Utilizing these basic schemes, this chapter demonstrates the methodology of applying control science to sustainable home design. This methodology is aimed to help not only promote sustainable design but also design sustainable homes.

The methodology is illustrated in the following two sections. The next section shows two-step preparatory work for sustainable home design. Utilizing these two steps, Section 3, which is short but significant, demonstrates the control system for promoting sustainable home design.

2. Two-step preparatory work for sustainable home design

As shown in the previous chapter, the preparatory work for sustainable design consists of the two steps: (1) determining the relationships between the standard human activities and sustainability and (2) sustainability checkup on human activities as an object. When the "home" is identified as a category of human activities, the two steps are (1) "determining the relationships between the standard home and sustainability" and (2) "sustainability checkup on a home as an object."

2.1. Determining the relationships between the standard home and sustainability

The first step aims to select important elements of the standard home and comprehensively determine the relationships between the selected elements and sustainability [1, 2].

In order to efficiently select important elements, we have examined two main factors, namely, "material" and "space" (**Table 1**) [1]. "Material" considers home as the aggregate of material elements, including framework, exterior, interior, and piping. "Space" regards home as the aggregate of spatial elements, including rooms and areas [1]. On the basis of these two factors, we have selected important elements, as shown in the left column of **Table 2**. "Material elements" are from "framework" to "outdoor facilities"; "spatial elements" are from "total floor" to "garden area."

We have subsequently determined the relationships between these elements and internal stability and fundamental stability (**Table 2**). That is to say, examining the relationships between each element and the stability conditions, we have identified variables that indicate the degree of stability [1, 2]. Moreover, we have set these variables' desired values that can satisfy relevant stability conditions [1, 2].

In addition, **Table 2** is the first updated version, which has been revised because of several reasons. First, we have revised the table so that following it leads to long-life quality housing (LQH, *Choki Yuryo Jutaku*) certification. The LQH certification began in 2009, and after that it has rapidly spread in Japan due to a variety of incentives, including tax reduction [3]. The second reason is the addition of adaptation measures against impacts resulting from climate change. Moreover, we have taken accessible and universal design more extensively, adding several spatial elements, such as stairs, hallway, and main access route to the entrance. Other reasons are wide spread of new energy-saving technology, namely, LED light, and minor changes in desired values and expressions.

Factor	Examples of elements (details)
Material	• Framework (pillar, beam, etc.) • Exterior (outer wall, roof, etc.) • Interior (floor, inner wall, ceiling, etc.) • Piping (water pipe, drainage pipe, gas pipe, etc.)
Space	• Room (living room, bedroom, dining room, kitchen, bathroom, etc.) • Area (exterior area, garden area, etc.)

Table 1. Two factors on selecting important elements of home [1].

Element	Variable	Desired value	Stability condition
Framework	Resistance to earthquakes	JHPIS 1.1: Grade 2 or over	• Safety
	Durability	JHPIS 3.1: Grade 3	• Sustainable resources
	Materials	CASBEE LR_H2 1.1: Level 4 or over	• Sustainable resources
Exterior (outer wall, roof, etc.)	Fire resistance (outer wall)	JHPIS 2.6: Grade 3 or over	• Safety
	Shape and color	Consideration for the landscape	• Health
	Durability	CASBEE Q_H2 1.2 and 1.3: Level 4 or over	• Sustainable resources
	Materials	CASBEE LR_H2 1.3: Level 4 or over	• Sustainable resources
Thermal insulation	Thermal insulation performance	JHPIS 5.1: Grade 4	• Health • Enviro-preservation • Sustainable resources
Windows and doors	Thermal insulation performance	JHPIS 5.1: Grade 4	• Health • Enviro-preservation • Sustainable resources
	Sunlight adjustment capability	CASBEE Q_H1 1.1.2: Level 4 or over	• Health • Enviro-preservation • Sustainable resources
	Sound insulation performance	CASBEE Q_H1 4: Level 4 or over	• Health
	Measures to prevent intrusions	CASBEE Q_H1 2.3: Level 4 or over	• Safety
	Protection of glass against impacts	With shutters	• Safety
Interior	Measures against formaldehyde	CASBEE Q_H1 2.1: Level 5	• Health
	Materials	CASBEE LR_H2 1.4: Level 4 or over	• Sustainable resources
Bathtub	Heat insulation	Insulated	• Enviro-preservation
Piping	Measures for maintenance	JHPIS 4.1: Grade 3	• Sustainable resources
	Method of water and hot-water piping	Header and pipe-in-pipe system	• Enviro-preservation • Sustainable resources
Water heater	Type of water heater	CASBEE LR_H1 2.2.1: Level 5	• Enviro-preservation • Sustainable resources
Appliances	Energy-saving standard achievement rate	100% or more (three or more stars)	• Enviro-preservation • Sustainable resources
Lighting fixtures	Type of light	LED	• Enviro-preservation • Sustainable resources
Equipment for harnessing natural energy	Harnessed natural energy	100% or more of the total energy usage	• Health (in crises) • Safety (in crises) • Enviro-preservation • Sustainable resources
Equipment for rainwater use	Rainwater equipment	CASBEE LR_H1 3.2: Level 4 or over	• Health (in crises) • Safety (in crises) • Enviro-preservation • Sustainable resources

Element	Variable	Desired value	Stability condition
Water-using equipment	Water-saving functions	CASBEE LR_H1 3.1: Level 4 or over	• Enviro-preservation • Sustainable resources
Outdoor facilities (fence, etc.)	Form	Not blocking sightlines	• Safety • Mutual help
	Appearance	Consideration for the landscape	• Health
	Materials	CASBEE LR_H2 1.5: Level 5	• Sustainable resources
Total floor	Total floor area	75m² or more [Note 3]	• Health
Specified bedroom	Routes to toilet and bath area, dining room, kitchen, and entrance	Accessible without steps	• Health • Safety
	Internal floor space	9 m² or more	
Areas relating to water use and hot-water supply	Areas in the home	Placing them closer	• Enviro-preservation • Sustainable resources
Position and area of windows	Natural ventilation	CASBEE Q_H1 1.2.1: Level 5	• Health • Enviro-preservation • Sustainable resources
	Ratio of total window area to floor area in each living space	20% or more	
Toilet	Internal length or spacing	JHPIS 9.1: Grade 3 or over	• Health • Safety
	Handrails which help users sit and stand	Installed	
Bathroom	Floor space and width	JHPIS 9.1: Grade 3 or over	• Health • Safety
	Handrails help users go in and out of the bathtub	Installed	
Stairs	Grade of steepness	JHPIS 9.1: Grade 3 or over	• Health • Safety
	Handrails	Installed	
Doorways	Differences in level	No differences	• Health • Safety
	Width	75 cm or more (bath, 60 cm or more)	
Hallway	Width	78 cm or more (pinch points, 75 cm or more)	• Health • Safety
Main access route to the entrance	Surface	Level or sloping	• Health • Safety
	Width	90 cm or more	
Slope	Grade of steepness	1/8 or less	• Health • Safety
	Handrails	Installed	
Garden area	Ratio of the garden area to the exterior area	40% or more	• Enviro-preservation

[Notes] (1) JHPIS stands for the *Japan Housing Performance Indication Standards (for new homes)*; (2) CASBEE stands for *CASBEE for Detached Houses (New Construction) – Technical Manual 2010 Edition*; (3) At least one story's area (excluding stairs) is 40 m² or more.

Table 2. Relationships between the standard home and sustainability.

The rest of this section concisely describes the relationships between each material or spatial element and sustainability, in order from the top of **Table 2**.

2.1.1. Material elements

- Framework

Considering the relationship between "framework" and "safety," a condition of internal stability, we have selected "resistance to earthquakes" as a variable and set its desired value at "Grade 2 or over" in the "seismic resistance grades (prevention of collapse of building structures)" of JHPIS, that is, the Japan Housing Performance Indication Standards (for new homes) [4]. The LQH certification requires satisfying "Grade 2 or over" likewise [5]. "Grade 2" means that the building can withstand 1.25 times the strength of an earthquake stipulated in the Building Standards Act of Japan [4]. In Japan, the strength of framework against earthquakes is regarded as extremely important since Japan is a major quake-prone country.

Furthermore, in areas of strong wind or heavy snowfall, "resistance to wind" or "resistance to snow load" needs to be included as a variable, although both of which are excluded from the table.

On the other hand, examining the relationship between "framework" and a condition for fundamental stability, namely, "sustainable use of natural resources," we have identified "durability" and "materials" as variables [2]. Moreover, we have set the desired value of "durability" at "Grade 3" in the "Deterioration resistance grades (Building frames, etc.)" of JHPIS [2]. "Grade 3" requires measures to extend the period of time between the construction and the first large-scale renovation up to three generations (about 75–90 years) or more, under normally assumed natural conditions and maintenance [4]. The LQH certification also requires securing this target grade [5].

Meanwhile, we have set the desired value of "materials" at "Level 4 or over" in the assessment levels of the "Use of resource-saving materials and less waste-producing materials" of CASBEE, namely, *CASBEE for Detached Houses (New Construction) – Technical Manual 2010 Edition* [2]. In the case of a wooden house, for example, "Level 4" requires that wood from sustainable forests is used for more than half of the building frames [6].

- Exterior (outer wall, roof, etc.)

As for "exterior," which includes roofs and outer walls, we have identified "fire resistance" and "shape and color" as variables relating to internal stability. The desired value of outer walls' "fire resistance" has been set at "Grade 3" in the "fire resistance grades" of JHPIS. "Grade 3" requires that flames are blocked for 45 min or more [4]. Meanwhile, the "shape and color" of the exterior requires "consideration for the landscape" as its desired value, so as to improve scenery or facilitate harmony with the surrounding landscape.

On the other hand, we have selected "durability" and "materials" as variables relating to fundamental stability. The desired value of "durability" is set at "Level 4 or over" in the assessment levels of the "Exterior wall materials" and "Roofing materials" of CASBEE. "Level 4" requires that a service life of 50 years to less than 100 years can be expected [6]. In the case of a service life of less than 50 years, however, the assessment levels can be raised if "ease of replacement" or "deterioration mitigation treatment" is considered [6]. Meanwhile, the desired value of "materials" is set at "Level 4 or over" in the assessment levels of the "Exterior materials" of CASBEE. "Level 4 or over" requires higher-level efforts in utilization of materials which promote resource saving or waste prevention such as recycled, renewable, and recyclable materials [6].

- Thermal insulation

We have identified "thermal insulation performance" as the variable of "thermal insulation" and set its desired value at "Grade 4" in the "Energy-saving action grades (Thermal insulation performance grades)" of JHPIS, which is the highest in the grades [4]. "Grade 4" requires that measures are taken to reduce energy use to a significant degree [4], details of which are shown in the *Judgement Criteria on Improvement in Housing Energy Efficiency for Building Owners*. According to this judgment, thermal insulation performance of the building was mainly evaluated, based on "thermal loss coefficient (Q)" (**Table 3**). The standard value of "Q" varies, depending on climate classification; for example, that of the classified area where it is relatively warm (including Tokyo) has been set at "2.7 $W/(m^2 * K)$ or less" [7]. Recently, this judgment criterion has been revised into the new version. According to the current judgment criteria, thermal insulation performance is evaluated on the basis of "building envelope's average heat transmission coefficient (U_A)" (**Table 3**). The criterion of "U_A" for the area where Tokyo is included has been set at "0.87 $W/(m^2 * K)$ or less" [8]. In addition, the *Guidelines for Design, Construction, and Maintenance on Improvement in Housing Energy Efficiency* show details of the criteria, including standard values by building parts or construction materials as well as technical guidelines for meeting the criteria [9].

"Thermal insulation performance" is important because it is related to both internal stability and fundamental stability. An increase in thermal insulation performance contributes to the sustainable use of natural resources and environmental preservation through a reduction in energy usage for heating and air-conditioning.

Meanwhile, there have been many studies which show correlation between higher thermal insulation performance and residents' better health. For example, empirical research in New Zealand has demonstrated that insulating existing homes led to significant improvements in

	Main housing thermal performance criterion stipulated in the judgment criteria for building owners	**Standard value in Tokyo area**
Former criterion	Thermal loss coefficient of the building (Q) $Q\,[W/(m^2 K)] = \dfrac{\text{thermal loss from the building}}{(\text{total floor area}) \times (\text{temperature difference unit})}$	$Q = 2.7\ [W/(m^2 K)]$ or less
Current criterion (2013~)	Building envelope's average heat transmission coefficient (U_A) $U_A[W/(m^2 K)] = \dfrac{\text{thermal loss from the building}}{(\text{total area of building envelope}) \times (\text{temperature difference unit})}$	$U_A = 0.87\ [W/(m^2 K)]$ or less

Table 3. Main housing thermal performance criteria stipulated in the judgment criteria for building owners of Japan.

the residents' self-reported health and in taking days off from school and work [10]. Similarly, research in Japan has shown that upgrading of thermal insulation performance decreased the occupants' prevalence rates of various diseases, such as allergic rhinitis, bronchial asthma, atopic dermatitis, and heart diseases [11]. Furthermore, a recent survey in Japan has indicated that improvement in thermal insulation performance increases indoor temperature in winter and reduces the occupants' blood pressure [12]. Reduction of blood pressure leads to decrease in the risk of heart attack and stroke.

- Windows and doors

We have selected five items as the variables of "windows and doors," that is, thermal insulation performance, sunlight adjustment capability, sound insulation performance, measures to prevent intrusions, and protection of glass against impacts.

Higher "thermal insulation performance" of openings is essential for stable indoor temperature and energy conservation. The ratio of heat flow through openings including windows is much larger than other parts such as walls and floors. According to an analysis in Japan, heat flow through openings accounts for more than half of the total heat flow: 58% of the total outflow while heating in the winter and 73% of the total inflow while air-conditioning in the summer [13]. As described in the above section of "thermal insulation," higher thermal insulation performance helps occupants' better health, as well as environmental preservation and sustainable use of natural resources. We have set the desired value at the highest "Grade 4" of the "energy-action grades (thermal insulation performance grades)" of JHPIS.

"Sunlight adjustment capability" evaluates the design of windows that blocks solar radiation during summer and captures it during winter, based on the sunlight penetration ratio. Higher sunlight adjustment capability also contributes to residents' better health, in addition to fundamental stability. The desired value is set at "Level 4 or over" of the relevant item of CASBEE. "Level 4" requires that the building can reduce the sunlight penetration ratio in the subject windows to 0.45 or less in the summer. "Level 5," the highest level, requires reducing it to 0.3 or less in the summer and 0.6 or more in the winter [6]. The factors that have influence on the sunlight penetration ratio are (a) type of glass, (b) solar shading materials such as lace curtains and blinds, and (c) eaves [6]. In addition, deciduous trees that create shade over almost the entire surface of the target window during the summer can be in the calculation as a factor of solar shading equivalent to eaves [6]. In this way, in order to meet sunlight adjustment capability, related elements such as curtains, blinds, eaves, and even trees are often required to work together.

Higher "sound insulation performance" of windows and doors is essential for indoor quietness against outdoor noise. The desired value of this variable has been set at "Level 4 or over" of the "Quietness" of CASBEE. "Level 4" and "Level 5" of the "Quietness" correspond to "Grade 2" and "Grade 3" of the "Transmission loss grades (Exterior wall openings)" of JHPIS, respectively [6]. "Grade 2" requires "equal to or higher than Rm (1/3) – 20db," which means 20db or more of sound transmission loss, measured by one-third octave band analysis. Similarly, "Grade 3" requires "equal to or higher than Rm (1/3) – 25db" [4].

Considering the relationship with safety, we have identified "measures to prevent intrusions" as a variable and set its desired value at "Level 4 or over" in the assessment levels of the "Precautions against crime" of CASBEE. "Level 4" requires that, regarding openings whose sizes have a risk of intrusion, effective measures to prevent intrusion have been taken for the entrance to the building and other openings whose lower edge is 2 m or less from ground level [6]. In the above explanation, "effective measures" include the installation of two or more locks in different places and attachment of covers such as shutters [6].

According to the IPCC, climate change is projected to increase impacts from extreme weather events, such as heat waves, droughts, floods, cyclones, and wildfires. A first step toward adaptation to future climate change is reducing vulnerability and exposure to present climate variability [14]. Therefore, it must be significant to protect the most vulnerable part of housing exterior, that is, window glass. Based on the above recognition, we have added "protection of glass against impacts" as a variable and set its desired value as "with shutters." Shutters are expected to reduce risks from fires, typhoons, tornadoes, and flying objects. In addition, *Your Home: Australia's guide to environmentally sustainable homes* also recommends installation of window protection like shutters, as an adaptation strategy to prepare against impacts such as bushfires, cyclones, and thunderstorms [15].

- Interior

"Interior," which includes floors, inner walls, and ceilings, requires "measures against formaldehyde" and "materials" as its variables.

Considering the relationship with "health," a condition of internal stability, we have identified "measures against formaldehyde" as a variable of interior. Formaldehyde is a colorless, flammable gas at room temperature and has a strong odor [16]. Formaldehyde is used in making building materials and many household products; for example, it is found in pressed-wood products, glues, permanent press fabrics, paper product coatings, and certain insulation materials [16]. Exposure to formaldehyde can cause adverse health effects, such as irritation of the skin, eyes, nose, and throat; high levels of exposure may cause some types of cancers [16]. We have set the desired value of "measures against formaldehyde" at "Level 5" of the "Countermeasures against chemical contaminants" of CASBEE. "Level 5" is equivalent to "Grade 3" in the section of the JHPIS's "Countermeasures against formaldehyde (Interior, ceiling plenum, etc.)" [6]. "Grade 3," the highest grade, means that formaldehyde emissions from interior finish and base materials are "extremely low." This requires the use of "F-four-star" certified products, which are the top-rated products in the formaldehyde emission standards according to the Japanese Industrial Standards and Japanese Agricultural Standards [4, 6].

On the other hand, we have set the desired value of "materials" as "Level 4 or over" of the CASBEE's relevant assessment item, "Interior materials." "Level 4 or over" requires higher-level efforts in utilization of materials which promote resource saving or waste prevention such as recycled, renewable, and recyclable materials [6].

- Bathtub

We have attached importance to "heat insulation" as a variable of the "bathtub" since "insulated" bathtubs can reduce heat loss of the hot water. This consideration is necessary because

http://dx.doi.org/10.5772/intechopen.71324

of a Japanese lifestyle; people frequently take a bath and usually share the same hot water in the bathtub with their family members.

- Piping

"Piping," including drainage pipes, water pipes, hot-water pipes, and gas pipes, needs "measures for maintenance" as an important variable toward a long service life. The desired value of the variable has been set at "Grade 3" of the "Maintenance grades" of JHPIS. Grade 3 requires consideration for making maintenance easier, such as not burying piping under concrete and creating openings for cleaning and inspection [4].

In addition, we have selected "method of water and hot-water piping" as another variable of piping and set its desired value at the "header and pipe-in-pipe system." In this piping system, water supply and hot-water supply branch into water pipes and hot-water pipes at the header, as shown in **Figure 1(A)**. Each water pipe or hot-water pipe connects the header and each faucet without any joints. Meanwhile, the pipes used in this system have double-tube structure. The outer plastic pipes play a role of guide and protection of the inner plastic pipes, which are usually made of cross-linked polyethylene [17].

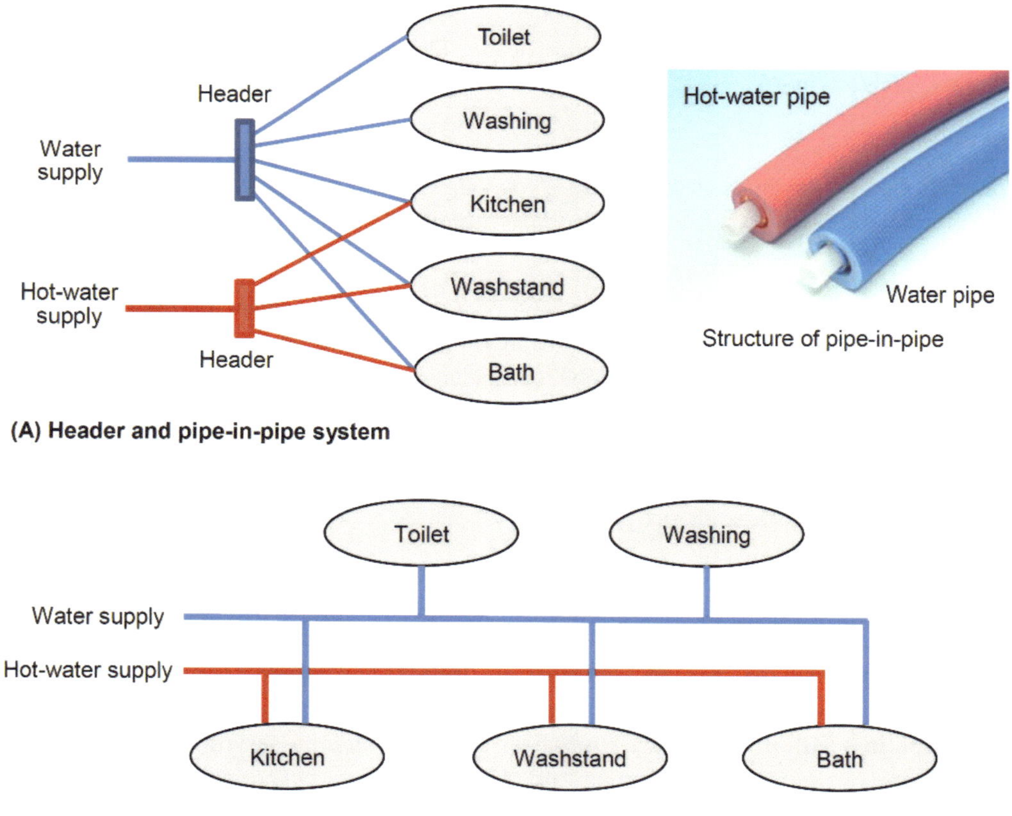

Figure 1. Header and pipe-in-pipe system and branch piping system.

As compared with the conventional "branch piping system," the schematic depiction of which is demonstrated in **Figure 1(B)**, the "header and pipe-in-pipe system" has various advantages. First, this system is superior in durability and maintenance due to the following reasons: (1) unlike conventional metal pipes, plastic pipes do not corrode, and (2) replacement of the inner pipes is easy because both inner and outer pipes are flexible and jointless between the header and each faucet [17, 18]. Second, this piping system is more energy saving since the diameter of the hot-water pipes is normally smaller than that in the branching system, and therefore the wastage of hot water can be reduced [6]. In addition, the piping work of this system is easier than the conventional method, and the installation time can be reduced [17, 18]. Furthermore, the flow of water or hot water is stable even if more than two faucets are used at the same time [17, 18]. As a result of these advantages, this "header and pipe-in-pipe system" has been becoming widespread in Japan since around 1990 [17, 18].

- Water heater

We have identified "type of water heater" as a key variable of the "water heater." The desired value of the type of water heater has been set at "Level 5," the highest level, in the hot-water supply equipment assessment levels of CASBEE. This level includes most energy-efficient types of water heaters, that is, (1) fuel burning, latent-heat recovery, instant-supply-type water heater, (2) electric heat-pump water heater, (3) solar water heater, and (4) solar hot-water supply system [6].

- Appliances

Home appliances are necessary to be energy-saving devices. We have identified the variable of such appliances as "energy-saving standard achievement rate" and set its desired value at "100% or more," in principle.

As demonstrated in **Figure 2**, an energy-saving standard achievement rate is displayed on "energy-saving labels," with the green or orange mark and approximately annual electricity consumption. The green mark is the symbol of achievement, which means the product's energy-saving standard achievement rate is 100% or more. The orange mark is the symbol of nonachievement, which means the rate is less than 100% [19]. In addition, the standard of energy-saving standard achievement rate is determined, based on the energy-saving level of the most energy-efficient products in each appliance [20].

Meanwhile, "unified energy-saving labels," an example of which is shown in the right of **Figure 2**, are used for several kinds of electrical appliances. On a unified energy-saving label, the energy-saving rating of the product is largely displayed, on a scale of one to five stars. In the case of appliances subject to unified energy-saving labels, we have set "three or more stars" as the desired value, following the manual of CASBEE for Detached Houses [6]. In addition, the target appliances of unified energy-saving labels include air-conditioners, refrigerators, televisions, and electric toilet seats. These electrical appliances consume much energy, and there are large differences in energy-saving performance between products; therefore, they have been selected as the target appliances of the unified energy-saving label [21].

• Lighting fixtures

We have identified the variable of "lighting fixtures" as "type of light" and set its desire value at "LED." Main reasons why this revised version has restricted only LED lights and excluded fluorescent lights are as follows: (1) superior energy-saving efficiency, (2) significantly longer lifespan, and (3) recent price reduction and rapid spread of LED technology [22].

• Equipment for harnessing natural energy

We have identified "harnessed natural energy" as the variable of "equipment for harnessing natural energy" such as solar panels and set at "100% or more of the total energy usage" as its desired value. This desired value means achieving net zero energy or energy plus housing. In addition to environmental preservation and sustainable use of natural resources, equipment for harnessing natural energy also contributes to health and safety in crises by generating emergency energy.

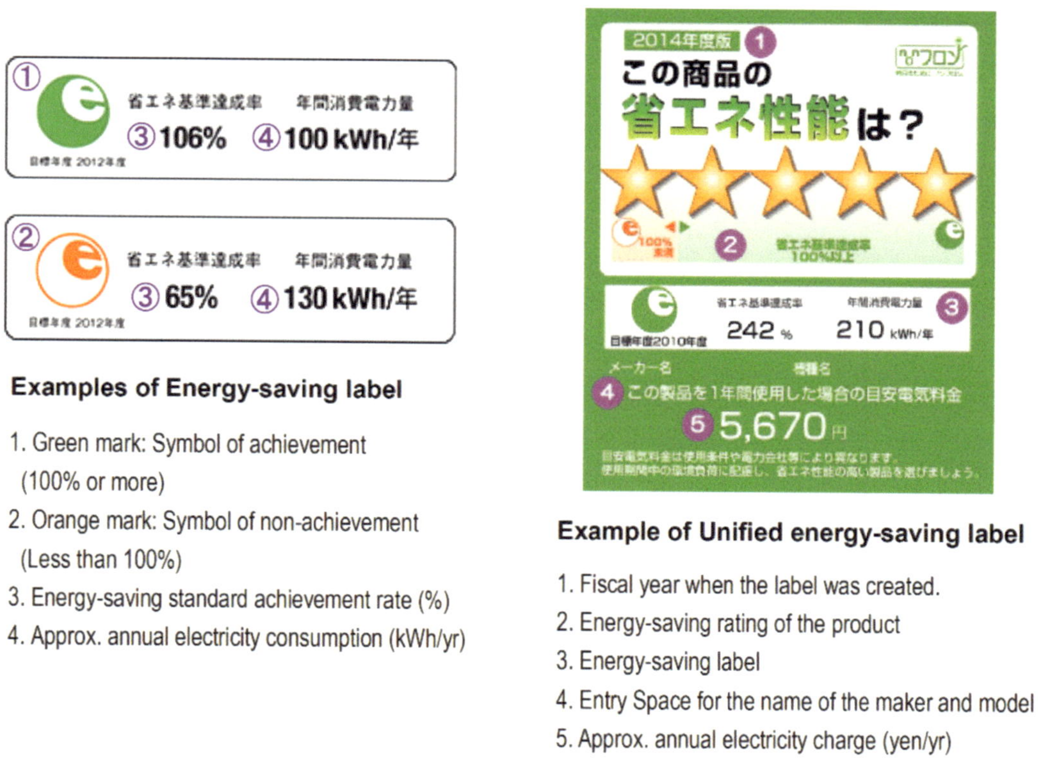

Figure 2. Examples of energy-saving label and unified energy-saving label [19].

- Equipment for rainwater use

If "equipment for rainwater use" is installed, it can reduce the quantity of water supply and contributes to fundamental stability. Reducing water supply leads to energy conservation and reductions in CO_2 emissions because energy is consumed through the process of water purification and distribution. Moreover, storing rainwater also contributes to health and safety in crises, by securing emergency water.

We have identified "rainwater equipment" as the variable and "Level 4 or over" of the CASBEE's relevant item as its desired value. "Level 4" requires installing a rainwater tank with a capacity of 80 liters or more. "Level 5" requires installing a rainwater utilization system with a cleaning water function for indoor use such as toilet flushing [6].

- Water-using equipment

"Water-using equipment," including toilet bowls, faucets, and shower heads, requires "water-saving functions" as its key variable. The desired value of this variable has been set at "Level 4 or over" of the CASBEE's relevant item, "Water-saving systems." In order to satisfy this desired value, home designers must adopt two or more water-saving efforts from the following five efforts: (1) water-saving-type toilets, (2) kitchen water-saving-type faucets, (3) bath water-saving-type faucet (4) dish washer, and (5) other water-saving methods [6].

- Outdoor facilities (fence, etc.)

We have selected "form" and "appearance" as variables relating to internal stability. "Form" of outdoor facilities, especially fences and barriers, should be "not blocking sightlines" since good visibility can bring "safety" and "mutual help" through preventing crime and allowing face-to-face communication. Meanwhile, "appearance" of outdoor facilities, such as shape and color, needs "consideration for the landscape," so as to improve scenery or promote harmony with the surrounding.

On the other hand, considering the relationship between "outdoor facilities" and "sustainable use of natural resources," we have identified "materials" as a variable and "Level 5" of the relevant item of CASBEE as its desired value. Level 5 requires using any of the following materials: (1) recycled materials, (2) reused materials, (3) wood produced from sustainable forests, and (4) natural materials which quickly become usable such as bamboo [6]. In addition, a note of this section says "recyclable materials" such as aluminum are acceptable, although this is excluded from the list [6].

2.1.2. Spatial elements

- Total floor

We have added "total floor" to the list of spatial elements because the long-life quality housing (LQH) certification requires the satisfactory size of the "total floor area." To be concrete, the

LQH certification requires "75m^2 or more" as the criterion of the total floor area of detached houses. The criterion adds the proviso that at least one story's floor area (excluding stairs) is 40m^2 or more [5]. In addition, "75m^2 or more" is equivalent to the floor area for two-person households of the "general-type target housing floor area level," which has been provided in the "Basic Plan for Housing (National Plan)" of Japan [6].

- Specified bedroom

A "specified bedroom" means a bedroom which is used or expected to be used by elderly or wheelchair users [23]. "Routes to the toilet, bath, dining room, kitchen and entrance" from the specified bedroom should be "accessible without steps." Therefore, all of such essential rooms and areas need to be arranged on the same floor, unless the house is equipped with an elevator or a lift.

Moreover, a specified bedroom requires "9 m^2 or more" as its "internal floor space." We have set this desired value on the basis of "Grade 3" of the JHPIS's relevant item, "Elderly friendliness grades (Dedicated spaces)" [23].

- Areas relating to water use and hot-water supply

"Areas relating to water use and hot-water supply" means a wet area (kitchen and bathroom area) and the area of a water heater. If such "areas in the home" are placed closer, the total length of water and hot-water piping and drainage piping can be reduced. Moreover, this consideration helps to reduce heat loss from hot-water piping.

- Position and area of windows

When planning "position and area of windows," we need to consider "natural ventilation" and "daylighting," both of which relate to fundamental stability and internal stability.

We have set the desired value of "natural ventilation" at "Level 5" of the CASBEE's relevant item, "Allowing breezes in and heat out." "Level 5" requires that the house has windows facing two or more directions in all living spaces [6]. Even if there is a window facing one direction, the house may be rated as "Level 5" if it is designed to promote ventilation and heat removal. Such design methods include securing paths for ventilation throughout the house, for example, by using sliding doors or latticed doors [6]. Adequate natural ventilation helps to reduce energy for air-conditioning as well as make the indoor environment more comfortable and healthier.

Meanwhile, we have identified "ratio of total window area to floor area in each living space" as the variable relating to daylighting and set its desired value at "20% or more." The value "20%" exceeds the legally stipulated value, or 1/7 (14.3%), and is equivalent to the satisfactory level of the relevant assessment item of CASBEE [6]. Taking in daylight through windows gives a sense of spaciousness to the occupants [24]. Moreover, recent studies show that bathing in daylight normalizes our biorhythm and contributes to health, for example, by improving sleep disorder, depression, and cognitive function; for example, see [24, 25]. On the other hand, the use of daylight leads to reducing energy for illumination.

- Toilet

We have selected "internal length or spacing" and "handrails which help users sit and stand" as variables of a "toilet." We have set the desired value of "internal length or spacing" at "Grade 3" of the JHPIS's relevant item, "Elderly friendliness grades (Dedicated spaces)," as the minimum level. "Grade 3" requires to meet either of the following two conditions: (1) at least 130 cm as the internal length of the space or (2) at least 50 cm as a spacing from the front rim or side rim of the toilet bowl [23]. Moreover, "handrails which help users sit and stand" need to be "installed."

- Bathroom

In Japanese homes, a "bathroom" is usually arranged separately from a toilet. Similar to a toilet, we have identified "floor space and width" and "handrails which help users go in and out of the bathtub" as variables of a "bathroom." We have set the desired value of "floor space and width" at "Grade 3" of the JHPIS's relevant item. "Grade 3" requires to satisfy both of the following two criteria: (1) at least 130 cm as the internal width of the space and (2) at least 2.0 m^2 as the internal floor space [23]. Furthermore, "handrails which help users go in and out of the bathtub" are necessary to be "installed."

- Stairs

It is essential to improve the safety of "stairs," in order to prevent accidental falls. We have identified "grade of steepness" and "handrails" as variables of stairs. We have set the desired value of "grade of steepness" at "Grade 3" of the JHPIS's relevant item. "Grade 3" requires to satisfy all of the following three criteria: (1) grade of steepness = rise/run ≤22/21, (2) 550 mm ≤ (rise * 2 + run) ≤ 650 mm, and (3) run ≥195 mm [23]. Meanwhile, "handrails" need to be "installed" at least on one side.

- Doorways

A "doorway" is a space where a door opens and closes. No "differences in level" in doorways allow everyone including elderly, children, and wheelchair users to pass through smoothly. Meanwhile, we have set the desired value of the "width" of doorways at "75 cm or more" and that of a bathroom's doorway at "60 cm or more." These desired values correspond to the standard values provided in "Grade 3" of the JHPIS's relevant item, "Elderly friendliness grades (Dedicated spaces)" [23].

- Hallway

Similar to doorways, we have set the desired value of the "width" of a "hallway" at "78 cm or more." The width of a hallway can be reduced to "75 cm or more" at pinch points such as beside a pillar. These figures are equivalent to the standard values shown in "Grade 3" of the JHPIS's relevant item [23].

- Main access route to the entrance and slope

"Main access route to the entrance" is usually the paths to the entrance from the street and/or car parking space. We have identified "surface" and "width" as the variables of this element. Easy and safe access requires the surface to be "level or sloping." The width of the main access route to the entrance should be "90 cm or more."

Moreover, a "slope" should also be easy and safe to access. We have identified "grade of steepness" of slopes as a variable and set its desired value at "1/8 or less." The other variable "handrails" should be "installed," unless the slope is sufficiently gentle, namely, 1/20 or less.

In addition, we have added these elements, aiming to take universal design into housing exterior as well as interior. As described in the chapter of "Introduction," universal design principles require homes to "be easy to enter" as well as other features, such as "be easy to move around in" [26]. Foreign universal design guidelines for detached houses include descriptions of dwelling access; to be concrete, both of the *Lifetimes Homes* of the UK and the *Livable Housing Design* of Australia require an access route to the dwelling entrance to be level or gently sloping [27, 28].

Meanwhile, Japanese society appears to be unconcerned or apathetic about the accessibility of pathways to detached houses. In fact, access routes to the entrance of almost all Japanese houses have steps, as shown in **Figure 3**. Moreover, the JHPIS does not describe the accessibility

Figure 3. Main access route to the entrance of common houses in Japan.

of the exterior area of private houses, whereas it provides detailed information on the indoor accessibility. Japanese public housing accessible design guidelines state that access routes to houses should be suitable for walking and using wheelchairs [29]; however, it does not refer to concrete specifications. Nevertheless, I occasionally come across places in which handrails have necessarily been added to the step area (**Figure 3**, upper right) or steps have been converted into a slope after construction (**Figure 3**, lower right). It is obvious that the accessibility of pathways to detached houses is necessary in Japan, the fastest aging country in the world.

When specifying the "main access route to the entrance" and "slope," we have referred to foreign universal design guidelines for detached houses, such as the *Lifetime Homes*, and Japanese universal design guidelines for public and commercial facilities, such as the *Architectural Design Standards for Facilitating Mobility of the Elderly, Handicapped and Others*. Moreover, we have taken the smallness of Japanese housing lots into consideration. The maximum grade of the slope, namely, "1/8," is equivalent to the maximum grade for short slopes, specified in the *Architectural Design Standards for Facilitating Mobility of the Elderly, Handicapped and Others* [30]. The minimum width of the main access route, namely, "90 cm," is equal to the minimum width of relevant routes required in the *Lifetime Homes* of the UK [27].

- Garden area

A "garden area" is an area with plants, including trees, shrubs, grasses, herbs, and vegetables. An area with plants is more environmentally friendly than that covered with concrete or asphalt, due to various reasons, such as a higher level of biodiversity, healthier water cycle, and mitigation of heat island phenomenon.

The variable of a garden area has been identified as "ratio of the garden area to the exterior area," and its desired value has been set at "40% or more." The desired value, 40% or more of the garden area to the exterior area, corresponds to "level 4" in the assessment levels of the "Greening of the premises" of CASBEE [6]. In addition, the garden area includes any planted area not only on the ground but also on the roof.

2.2. Sustainability checkup on a home as an object

The second step is "sustainability checkup on a home as an object." The second step starts with the measurement or estimation of the aforementioned variables of a home as an object [1, 2]. Subsequently, the measured or estimated values are compared with the desired values, and the comparison results are assessed [1, 2]. **Table 4** shows an example of "sustainability checkup on a home as an object," which is equivalent to the results of a checkup on an existing home.

In this case, the checkup results have simply been assessed whether the variable reaches the desired value or not, that is, "OK" or "No." The variables that have been assessed as "No" need to be identified as "controlled variables." In addition, this home is naturally identified as a "controlled object" because it includes controlled variables.

Element	Variable	Measured or estimated value	Assess.	Desired value
Framework	Resistance to earthquakes	JHPIS 1.1: Grade 1	No	JHPIS 1.1: Grade 2 or over
	Durability	JHPIS 3.1: Grade 1	No	JHPIS 3.1: Grade 3
	Materials	CASBEE LR_H2 1.1: Level 4	OK	CASBEE LR_H2 1.1: Level 4 or over
Exterior (outer wall, roof, etc.)	Fire resistance (outer wall)	JHPIS 2.6: Grade 3	OK	JHPIS 2.6: Grade 3 or over
	Shape and color	Consideration for the landscape	OK	Consideration for the landscape
	Durability	CASBEE Q_H2 1.2 and 1.3: Level 2	No	CASBEE Q_H2 1.2 and 1.3: Level 4 or over
	Materials	CASBEE LR_H2 1.3: Level 3	No	CASBEE LR_H2 1.3: Level 4 or over
Thermal insulation	Thermal insulation performance	JHPIS 5.1: Grade 1	No	JHPIS 5.1: Grade 4
Windows and doors	Thermal insulation performance	JHPIS 5.1: Grade 1	No	JHPIS 5.1: Grade 4
	Sunlight adjustment capability	CASBEE Q_H1 1.1.2: Level 3	No	CASBEE Q_H1 1.1.2: Level 4 or over
	Sound insulation performance	CASBEE Q_H1 4: Level 3	No	CASBEE Q_H1 4: Level 4 or over
	Measures to prevent intrusions	CASBEE Q_H1 2.3: Level 3	No	CASBEE Q_H1 2.3: Level 4 or over
	Protection of glass against impacts	With shutters	OK	With shutters
Interior	Measures against formaldehyde	CASBEE Q_H1 2.1: Level 5	OK	CASBEE Q_H1 2.1: Level 5
	Materials	CASBEE LR_H2 1.4: Level 4	OK	CASBEE LR_H2 1.4: Level 4 or over
Bathtub	Heat insulation	Not insulated	No	Insulated
Piping	Measures for maintenance	JHPIS 4.1: Grade 3	OK	JHPIS 4.1: Grade 3
	Method of water and hot-water piping	Branch piping system	No	Header and pipe-in-pipe system
Water heater	Type of water heater	CASBEE LR_H1 2.2.1: Level 1	No	CASBEE LR_H1 2.2.1: Level 5
Appliances	Energy-saving standard achievement rate	70–95%	No	100% or more (three or more stars)
Lighting fixtures	Type of light	Fluorescent	No	LED
Equipment for harnessing natural energy	Harnessed natural energy	0 (zero)	No	100% or more of the total energy usage
Equipment for rainwater use	Rainwater equipment	No equipment	No	CASBEE LR_H1 3.2: Level 4 or over
Water-using equipment	Water-saving functions	CASBEE LR_H1 3.1: Level 1	No	CASBEE LR_H1 3.1: Level 4 or over

Element	Variable	Measured or estimated value	Assess.	Desired value
Outdoor facilities (fence, etc.)	Form	Not blocking sightlines	OK	Not blocking sightlines
	Appearance	Consideration for the landscape	OK	Consideration for the landscape
	Materials	CASBEE LR_H2 1.5: Level 5	OK	CASBEE LR_H2 1.5: Level 5
Total floor	Total floor area	116 m^2	OK	75m^2 or more [Note 4]
Specified bedroom	Routes to toilet and bath area, dining room, kitchen, and entrance	With steps	No	Accessible without steps
	Internal floor space	9.7 m^2	OK	9 m^2 or more
Areas relating to water use and hot-water supply	Areas in the home	Placing them closer	OK	Placing them closer
Position and area of windows	Natural ventilation	CASBEE Q_H1 1.2.1: Level 5	OK	CASBEE Q_H1 1.2.1: Level 5
	Ratio of total window area to floor area in each living space	20–22%	OK	20% or more
Toilet	Internal length or spacing	Internal length: 120 cm Spacing: 55 cm	OK	JHPIS 9.1: Grade 3 or over
	Handrails which help users sit and stand	Installed	OK	Installed
Bathroom	Floor space and width	Floor space: 2.6 m^2 Width: 160 cm	OK	JHPIS 9.1: Grade 3 or over
	Handrails help users go in and out of the bathtub	Not installed	No	Installed
Stairs	Grade of steepness	25/21	No	JHPIS 9.1: Grade 3 or over
	Handrails	Installed	OK	Installed
Doorways	Differences in level	With differences	No	No differences
	Width	60–70 cm	No	75 cm or more (bath, 60 cm or more)
Hallway	Width	78 cm	OK	78 cm or more
Main access route to the entrance	Surface	With steps	No	Level or sloping
	Width	140 cm	OK	90 cm or more
Slope	Grade of steepness	No slope	No	1/8 or less
	Handrails			Installed
Garden area	Ratio of garden area to exterior area	63%	OK	40% or more

(1) JHPIS stands for the *Japan Housing Performance Indication Standards (for new homes)*. (2) CASBEE stands for *CASBEE for Detached Houses (New Construction): Technical Manual 2010 Edition*. (3) When this checklist is used for the inspection or evaluation of existing homes, *JHPIS (for existing homes)* and *CASBEE for Detached Houses (Existing Building): Technical Manual 2011 Edition* need to be referred to, instead of the "for new homes" version and "New Construction" version, respectively. (4) At least one story's area (excluding stairs) is 40 m^2 or more.

Table 4. An example of sustainability checkup on a home as an object.

Here I view the checkup results, choosing several elements from **Table 4**. Concerning "framework," two of the three variables, "resistance to earthquakes" and "durability," have been assessed as "No," because they are lower than the desired values. "Protection of glass against impacts," a variable of "windows and doors," has been assessed as "OK" since almost all windows of this home are equipped "with shutters." The "type" of the "water heater" used in this home is an energy-wasteful gas heater. As a result, it has been estimated at Level 1 of the relevant item of CASBEE and hence assessed as "No." "Areas in the home," the variable of "areas relating to water use and hot-water supply," has been assessed as "OK" because such areas are placed closer in this house. "Surface," a variable of "main access route to the entrance," is "with steps;" therefore, it has been assessed as "No."

3. Control system for promoting sustainable home design

Utilizing the "basic control system for sustainability" and the "two-step preparatory work for sustainable home design," we have produced the "control system for promoting sustainable home design" [2].

First of all, as demonstrated in **Figure 4**, we have derived two practical functions from the two-step preparatory work, namely, the "sustainable design guidelines" from Step 1 and the "sustainability checklist" from Step 2, respectively [2].

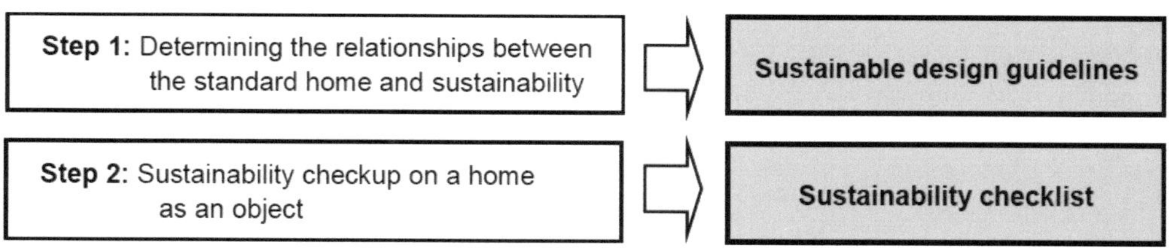

Figure 4. Two practical functions derived from the two-step preparatory work [2].

After that, we have formed the "control system for promoting sustainable home design" in which these two practical functions are incorporated [2]. **Figure 5** shows the block diagram of that control system. In this control system, "people involved in design" include homeowners, architects, designers, and homebuilders [2]. "Controlled objects" are both "new homes" and "existing homes" [2]. The following illustrates how to use the guidelines and checklist in the process of sustainable housing design, in the order of "new homes" and "existing homes."

3.1. New homes

When objects are new homes, first, information on the desired values reaches "people involved in design" through the "sustainable design guidelines" [2]. People involved make "drawings and specifications," so that the variables of home's elements can attain their desired values as much as possible [2]. At important steps in the design process, people involved in design check the drawings and specifications, by referring to the "sustainability checklist" [2]. After the construction is finished, the newly built home can be also evaluated against the "sustainability checklist" [2].

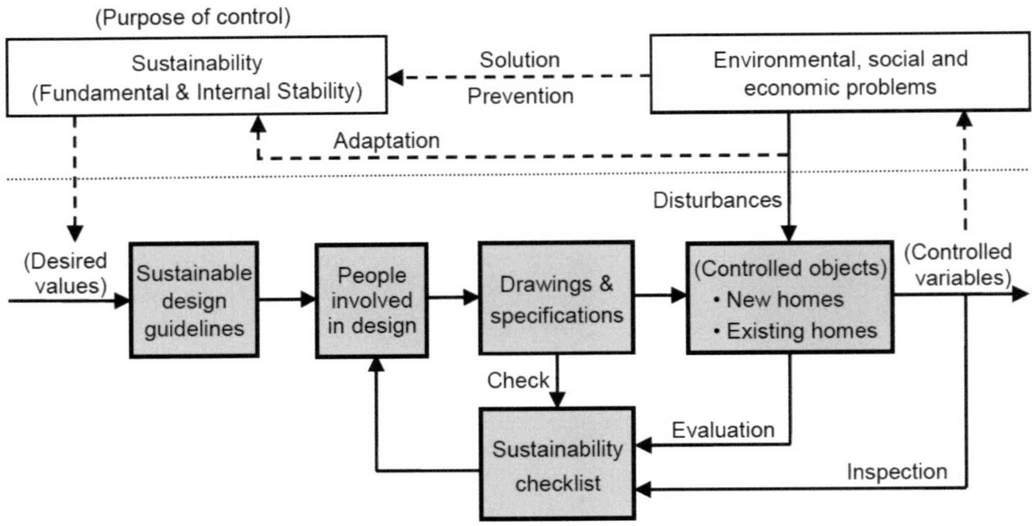

Figure 5. Control system for promoting sustainable home design [2].

3.2. Existing homes

When existing homes are the objects, the design process begins with "inspection" on the home as an object [2]. The "people involved in design" measure or estimate each element's variables of that home by referring to the "sustainability checklist" [2]. Next, they compare the measured or estimated variables with their desired values and assess the comparison results [2].

Table 4 in the previous section is equivalent to an instance of such inspection results. In addition, when inspecting an existing home and measuring or estimating variables by referring to CASBEE for Detached Houses or the JHPIS, "people involved in design" use the "Existing Building" version or "for existing homes" version, instead of the "New Construction" version or "for new homes" version, respectively [2]. "CASBEE for Detached Houses (Existing Building)" and the "JHPIS for existing homes" are almost the same as its new home version. However, the existing home version includes suitable assessment criteria for existing homes [2]. For example, as for "durability" of "exterior (outer wall, roof, etc.)," the existing home version of CASBEE shows the criteria for assessing the exterior's present condition and estimated remaining life at the assessment point of time [31].

After the inspection, the "people involved in design" usually make "drawings and specifications" for improvement, so that controlled variables satisfy their desired values as much as possible [2]. When "people involved" consider that improvement is technically difficult or costly, they can choose reconstruction instead of improvement [2]. Similar to the cases of new homes, they check the drawings and specifications for improvement or reconstruction against the "sustainability checklist [2]." Furthermore, sustainability of the actually improved or reconstructed homes can be evaluated against the checklist [2].

Author details

Kazutoshi Fujihira

Address all correspondence to: fujihira@kankyogaku.com

Institute of Environmentology, Tokyo, Japan

References

[1] Fujihira K. An approach to sustainable homes by applying control science. Advanced Materials Research. 2012;**2087-2092**:403-408. DOI: 10.4028/www.scientific.net/AMR.403-408.2087

[2] Fujihira K. System control for sustainability: Application to building design. In: Thomas C, editor. Complex Systems, Sustainability and Innovation. Rijeka, Croatia: InTech; 2016. p. 37-55. DOI: http://dx.doi.org/10.5772/65875

[3] Ministry of Land, Infrastructure, Transport and Tourism. Concerning Long-life Quality Housing Certification System (in Japanese) [Internet]. Available from: http://www.mlit.go.jp/common/001008009.pdf [Accessed: 2017-03-17]

[4] Ministry of Land, Infrastructure, Transport and Tourism. Japan Housing Performance Indication Standards (in Japanese) [Internet]. 2001. Available from: http://www.mlit.go.jp/jutakukentiku/house/torikumi/hinkaku/061001hyouji.pdf [Accessed: 2017-03-17]

[5] Center for Better Living, Association for Housing Performance Evaluation and Indication. Long-life Quality Housing Certification Manual (in Japanese) [Internet]. Available from: http://www.hyoukakyoukai.or.jp/kikan/pdf/chouki_manual.pdf [Accessed: 2017-03-17]

[6] Japan GreenBuild Council, Japan Sustainable Building Consortium. CASBEE for Detached Houses (New Construction)—Technical Manual. 2010 edition (in Japanese) ed. Tokyo, Japan: Institute for Building Environment and Energy Conservation; 2010. p. 178

[7] Ministry of Economy, Trade and Industry, Ministry of Land, Infrastructure, Transport and Tourism. Judgement Criteria on Improvement in Housing Energy Efficiency for Building Owners (in Japanese) [Internet]. 2009. Available from: http://www.mlit.go.jp/common/000038489.pdf [Accessed: 2017-03-17]

[8] Ministry of Economy, Trade and Industry, Ministry of Land, Infrastructure, Transport and Tourism. Judgement Criteria on Improvement in Energy Efficiency for Building Owners (in Japanese) [Internet]. 2013. Available from: http://www.mlit.go.jp/common/001082964.pdf [Accessed: 2017-03-17]

[9] Ministry of Land, Infrastructure, Transport and Tourism. Guidelines for Design, Construction, and Maintenance on Improvement in Housing Energy Efficiency (in Japanese)

[Internet]. 2013. Available from: http://www.mlit.go.jp/common/001013541.pdf [Accessed: 2017-03-17]

[10] Howden-Chapman P et al. Effect of insulating existing houses on health inequality: Cluster randomised study in the community. BMJ. 2007;**334**:460-464. DOI: http://dx.doi.org/10.1136/bmj.39070.573032.80

[11] Ikaga T, Eguchi R, Murakami S, Iwamae A, Hoshi T, Mizuishi T, Kawakubo S, Okumura K. Evaluation of investment in residential thermal insulation considering non-energy benefits delivered by health (in Japanese). Journal of Environmental Engineering (Transactions of AIJ). 2011;**76**:735-740. DOI: http://doi.org/10.3130/aije.76.735

[12] Ministry of Land, Infrastructure, Transport and Tourism. Interim Report of the Survey on Improvement in Housing Thermal Insulation Performance and its Effects on the Occupants' Health (in Japanese) [Internet]. Available from: http://www.mlit.go.jp/report/press/house07_hh_000164.html [Accessed: 2017-03-18]

[13] Japan Construction Material & Housing Equipment Industries Federation. Heat Flow Through Openings (in Japanese) [Internet]. Available from: http://www.kensankyo.org/syoene/qanda/mado/a_9.html [Accessed: 2017-03-17]

[14] IPCC. Climate Change 2014: Synthesis Report. Contribution of Working Groups I, II and III to the Fifth Assessment Report of the Intergovernmental Panel on Climate Change. IPCC: Geneva, Switzerland; 2014. p. 151

[15] Australian Government, Department of Climate Change and Energy Efficiency. Your Home: Australia's Guide to Environmentally Sustainable Homes, Adapting to Climate Change [Internet]. 2013. Available from: http://www.yourhome.gov.au/housing/adapting-climate-change [Accessed: 2017-03-17]

[16] United States Environmental Protection Agency. Facts About Formaldehyde [Internet]. Available from: https://www.epa.gov/formaldehyde/facts-about-formaldehyde [Accessed: 2017-03-17]

[17] Urban Renaissance Agency, Department of Technical Management. Development of Water Service Method by Using Resin Pipes: Header and Pipe-in-Pipe System (in Japanese) [Internet]. Available from: http://www.ur-net.go.jp/rd/corner-p/pdf/ur2006rd010-08.pdf [Accessed: 2017-03-19]

[18] Sato T, Sakato J, Sato T, Kubota T. Sugiyama J, Nakajima Y. Application of the header pipe-in-pipe system for sanitary installations of detached housing (in Japanese). Furukawa Electric News Reports. 2000;**105**:119-124.

[19] Association for Electric Home Appliances. Energy-Saving Labels and Unified Energy-Saving Labels (in Japanese) [Internet]. Available from: http://www.shouene-kaden2.net/learn/eco_label.html [Accessed: 2017-03-17]

[20] Energy Conservation Center, Japan. What is the Top-Runner Standard? (in Japanese) [Internet]. Available from: https://www.eccj.or.jp/machinery/toprunner/toprunner.pdf [Accessed: 2017-03-17]

[21] Energy Conservation Center, Japan. Unified Energy-Saving Labels (in Japanese) [Internet]. Available from: http://www.eccj.or.jp/machinery/labeling_program/03.pdf [Accessed: 2017-03-17]

[22] Carole Jacques Lux Research, Inc. LED Lighting Market to Grow Over 12-Fold to $25 Billion in 2023 [Internet]. Available from: http://finance.yahoo.com/news/led-lighting-market-grow-over-123000345.html;_ylt=AwrBEiEyM_hSeA4AIC7QtDMD [Accessed: 2017-03-17]

[23] Ministry of Land, Infrastructure, Transport and Tourism. Japan Housing Performance Indication Standards Evaluation Method Criteria (in Japanese) [Internet]. Available from: http://www.mlit.go.jp/common/000052960.pdf [Accessed: 2017-04-11]

[24] Japan Healthy Housing Association (Nihon Kenko Jutaku Kyokai), Department of Light and Visual Environment. Light for Healthy Housing Life (in Japanese) [Internet]. Available from: http://www.kjknpo.com/html_j/bukai/hikari16/jouhou/jouhou1.html [Accessed: 2017-03-18]

[25] Obayashi K et al. Positive effect of daylight exposure on nocturnal urinary melatonin excretion in the elderly: A cross-sectional analysis of the HEIJO-KYO study. The Journal of Clinical Endocrinology and Metabolism. 2012;**97**(11):4166-4173. DOI: 10.1210/jc.2012-1873

[26] Australian Government Department of Social Services. National Dialogue on Universal Housing Design—Strategic Plan [Internet]. 2011. Available from: https://www.dss.gov.au/our-responsibilities/disability-and-carers/program-services/government-international/national-disability-strategy-initiatives/livable-housing-design/national-dialogue-on-universal-housing-design-strategic-plan?HTML [Accessed: 2017-04-11]

[27] Habinteg Housing Association. Lifetime Homes Revised Criteria [Internet]. 2010. Available from: http://www.lifetimehomes.org.uk/data/files/For_Professionals/accessible_revisedlthstandard_final.pdf [Accessed: 2017-03-20]

[28] Livable Housing Australia. Livable Housing Design. 3rd edition [Internet] ed2013. Available from: http://livablehousingaustralia.org.au/library/SLLHA_GuidelinesJuly 2015-3.pdf [Accessed: 2017-03-20]

[29] Ministry of Land, Infrastructure, Transport and Tourism. Guidelines for Designing Houses Where Elderly People Live (in Japanese) [Internet]. Available from: http://www.pref.oita.jp/uploaded/attachment/119522.pdf

[30] Ministry of Land, Infrastructure, Transport and Tourism. Architectural Design Standards for Facilitating Mobility of the Elderly, Handicapped and Others (in Japanese) [Internet]. 2012. Available from: http://www.mlit.go.jp/jutakukentiku/build/barrier-free.files/guideline12.pdf [Accessed: 2017-03-13]

[31] Japan Green Build Council, Japan Sustainable Building Consortium. CASBEE for Detached Houses (Existing Building)—Technical Manual. 2011 edition (in Japanese) ed. Tokyo, Japan: Institute for Building Environment and Energy Conservation; 2011. p. 185

Chapter 5

Case Study: Detached House Designed by Following the Control System

Kazutoshi Fujihira

Additional information is available at the end of the chapter

http://dx.doi.org/10.5772/intechopen.71323

Abstract

The previous chapter has demonstrated the control system for promoting sustainable housing design in which the sustainable design guidelines and sustainability checklist are incorporated. Following this control system, we have actually designed and constructed a detached house. To be concrete, the homeowner and the architects of the housing manufacture have designed the home's parts, or elements, so that as much as possible the elements' variables meet their desired values. The sustainable design guidelines and sustainability checklist have been readily accepted because the material and spatial elements are equivalent to real parts of the home. After the home started to be used, we have obtained external evaluations of the home's sustainability performance. For example, CASBEE for Detached Houses, a comprehensive assessment system, has readily ranked the house in the highest "S." An energy-saving performance assessment has shown that this home has reduced energy consumption by over 70%, as compared with the average home. On the other hand, the reactions of the occupants and visitors have indicated the comfort, healthiness and safety of this house. Furthermore, this home has received a sustainable housing award, especially due to its extremely high sustainability and energy-saving performance.

Keywords: sustainable design guidelines, sustainability checklist, energy-saving performance, sustainable housing award

1. Introduction

The preceding chapter has shown the methodology of applying control system to sustainable housing design. Utilizing this methodology, we have conducted a case study. That is to say, following the control system for promoting sustainable home design, we have designed a home and constructed it. After the home began to be used, we have obtained evaluations of the home's environmental and sustainability performance.

The remaining three sections of this chapter illustrate the case study. The next section describes the design process of this house. Section 3 shows the results of the design and construction with many photos and diagrams. The last section demonstrates the environmental and sustainability performance of this house.

2. Design process

Building design process is divided into basic design and detailed design. In this case study, the homeowner, who is the author of this monograph, made the basic design by himself. Subsequently, the homeowner made a contract with a homebuilder whose skills appeared enough to satisfy the variables' desired values, which have been demonstrated in Table 2 of Chapter 4 [1]. After the contract, this homebuilder made the detailed design and constructed the house.

In the basic design stage, the homeowner used the "sustainable design guidelines" and "sustainability checklist," in addition to following related laws and regulations. To be concrete, he made the site plan, floor plans, and elevations, taking the information of relevant material and spatial elements into consideration. When arranging rooms of the house, first of all, he referred to two spatial elements: "specified bedroom" and "areas relating to water use and hot water supply." Subsequently, when determining the floor space of rooms and other areas, he considered spatial elements related to accessibility, such as "stairs," "specified bedroom," "toilet," and "bathroom." After that, the homeowner planned the elevation, considering several elements' variables. He determined "position and area of windows," taking daylighting and natural ventilation into consideration. He also considered variables of material elements related to basic plan, including sunlight adjustment capability of "windows," shape of "exterior," and energy harnessed by "equipment for harnessing natural energy." Furthermore, when making the exterior and garden plan, he mainly referred to three spatial elements: "garden area," "main access route to the entrance," and "slope."

When the detailed design stage began, the homeowner requested the designers of the homebuilder to refer to the "sustainable design guidelines" and "sustainability checklist." The designers readily accepted them since the material and spatial elements in the guidelines and checklist are equivalent to actual parts of the home [1]. Subsequently, examining the basic design results and various conditions, the designers and homeowner determined the site plan, floor plans, elevation, and fundamental specifications. Next, the designers of the homebuilder designed the home's elements, including framework, exterior, windows and doors, interior, and doorways, so that as much as possible, the elements' variables satisfy their desired values [1].

After the detailed design finished, we obtained building confirmation and Long-life Quality Housing (LQH) certification. After that, the home, a two-story wooden detached house, was built in a residential district on the outskirts of Tokyo.

3. Results of the design and construction

This section shows the results of the design and construction, in accordance with the elements and their variables.

First, **Figure 1** demonstrates the site plan and floor plans in which the descriptions of five spatial elements are included.

- Total floor

We have identified "total floor area" as a variable, following the certification criteria of the LQH. The LQH certification requires "75 m^2 or more" as the criterion of the total floor area of detached houses. This criterion adds the proviso that at least one story's floor area (excluding stairs) is "40 m^2 or more" [2].

The total floor area of this detached house is 122.96 m^2; both the areas of the house's first and second floor (excluding stairs) exceed 40 m^2. These figures sufficiently satisfy the desired value, which has led to the LQH certification.

- Specified bedroom

A "specified bedroom" means a bedroom which is used or expected to be used by elderly or wheelchair users [3]. When the owner couple moved into this house, they were in their late 50s and not disabled. However, in order to prepare for their future, we have considered this bedroom as a specified bedroom [1].

In this house, their bedroom and other essential areas for daily life, namely toilet, bath, dining room, kitchen and entrance, all have been placed on the first floor [1]. This floor planning enables them to have easy access to such areas without any steps [1].

Meanwhile, the "internal floor space" of this specified bedroom is 12.6 m^2, which has fulfilled the desired value, 9 m^2 or more.

- Areas relating to water use and hot water supply

"Areas relating to water use and hot water supply," namely bath, kitchen, washing room, toilet, and a place to put the water heater, have been placed closer [1]. This arrangement leads to reduction of heat loss from hot water piping. Moreover, this consideration reduces the total length of water and hot water piping and drainage piping [1].

- Position and area of windows

Considering natural ventilation and daylighting, we have determined "position and area of windows" of this home.

We have located windows so as to allow breezes in and heat out. As shown in **Figure 1** and **Table 1**, all of the four living spaces, namely living/dining room & kitchen, bedroom, multipurpose room, and study, have windows which face two or three directions. This arrangement of the windows meets the desired value, "Level 5" of the CASBEE's relevant item. In

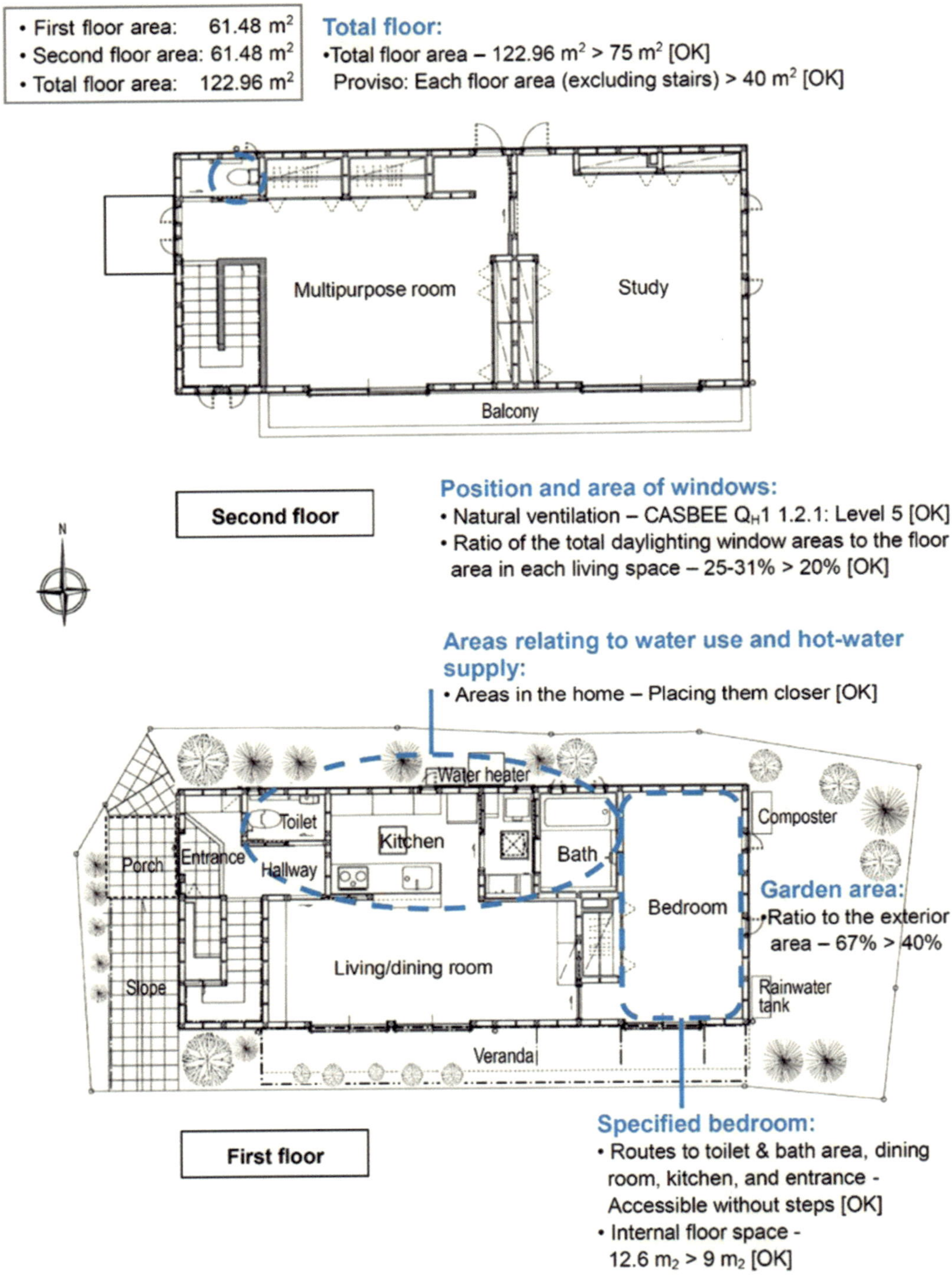

Figure 1. Site and floor plans.

Floor	Living space	Directions windows face (natural ventilation)	Ratio of the total window area to the floor area (daylighting) (%)
First	Living/dining and kitchen	South, North	28.5
	Bedroom	South, East	30.7
Second	Multipurpose room	South, North, West	25.6
	Study	South, East, North	26.2

Table 1. Natural ventilation and daylighting of the living spaces of this home.

addition, all of the inside doors are the sliding type, which enables the whole house easily to secure paths for natural ventilation.

In order to take sufficient daylight in, we have secured larger window areas, especially on the southern side of each living space. As demonstrated in **Figure 1** and **Table 1**, the ratios of the window area to the floor area in the four living spaces range between 25 and 31%. These values have substantially satisfied the desired value, 20% or more.

- Garden area

In this building site, the "garden area" surrounds the areas of the house, veranda, approach to the entrance, and subsidiary equipment such as the rainwater tank, as demonstrated in **Figures 1** and **2**. The garden area and the total exterior area are 45.9 and 68.7 m^2, respectively. Therefore, the "ratio of the garden area to the total exterior area" is 67%, which satisfies its desired value, 40% or more. In the garden area of this home, various kinds of trees, shrubs, and herbs have been planted. Moreover, in a part of the garden area near the rainwater tank and composter, several kinds of vegetables are cultivated every year.

Figure 3 shows external appearance, including panoramic views in winter and summer, with the descriptions of two material elements: "exterior" and "windows and doors."

- Exterior (outer wall, roof, etc.)

In order to restrain the spread of fire, fire resistance of outer walls and eaves soffit is important. The outer walls of this home, made of ceramic siding, can block flames for 60 minutes or more. Similarly, the eaves soffit can block flames for 45 minutes or more. These fire resistance capacities fulfill the desired value, "Grade 3" in the "Fire resistance grades" of JHPIS.

The landscape of this area includes disorder; therefore, we have strived to make it more attractive through the construction of this house. The "shape" is simple but accented by the balcony; the "color" of the outer walls, light beige, is coordinated with the colors of other parts such as the windows, entrance porch, and rainwater pipes. The solar panels have been placed not on roof materials but on sheathing roof boards so that the roof surface is almost flat. We are expecting that these considerations bring harmony and stability to the landscape.

Figure 2. Plants in the garden area.

The roof is almost entirely covered with solar panels; therefore, when evaluating the "durability" of the exterior, we have focused on outer walls. Durability of ceramic siding, this home's outer wall material, is expected to 40 years [4]. A service life of 40 years is assessed at "Level 3" in the assessment levels of the "Exterior wall materials" of CASBEE. However, this house uses an installation method that does not damage the building structure and window sashes when exterior wall materials are replaced. As a result, the assessment grade has been raised to "Level 4," which has met the desired value.

Meanwhile, we could not utilize "materials" which promote resource saving or waste prevention, such as recycled, renewable, and recyclable material, due to financial restrictions. As a result, the assessment level of this variable has fallen below the desired value.

- Windows and doors

All of the "windows" of this house are combined sash of aluminum and resin, which is high in both thermal insulation and strength, with low radiation double glazing. The thermal insulation performance of all large windows has been graded as "H-1," the highest in the classification of the Japanese Industrial Standards. Meanwhile, the entrance "door" is a metal flush door filled up with heat-insulating materials. The installation of the above windows and door has led to "$Q = 1.9\ W/(m^2 * K)$" as thermal loss coefficient of this house. This value has met the desired value of thermal loss coefficient in the area where Tokyo is included, namely "$Q = 2.7\ W/(m^2 * K)$ or less," which is equivalent to "Grade 4" in the "Energy-saving action grades (Thermal insulation performance grades)" of JHPIS [5].

Figure 3. External appearance.

Meanwhile, the "sunlight adjustment capability" of the windows facing south attains "Level 5" in the assessment levels of the CASBEE's relevant item [1]. "Level 5" requires that the sunlight penetration ratio of the subject windows can be adjusted to 0.6 or over in the winter and 0.3 or less in the summer [4]. The glass of the windows on the south side allows 63% of sunlight to penetrate to the rooms, which surpasses the desired value in winter, 60%. In the summer, in addition to the window glass and lace curtains, the balcony and the roof with pendent eaves can block the strong sunlight from entering the four living spaces, as shown in the left photo of **Figure 3**. The total reduction ratio amounts to 0.73; therefore, the sunlight penetration ratio in the summer can be reduced to 0.27. On the other hand, the deciduous tree that has been planted in front of the stairs' windows, blocks strong sunlight in the summer from entering the staircase. Accordingly, in this case, in addition to windows themselves, eaves, curtains, and even a tree work together, so as to control sunlight penetration.

The "sound insulation performance" of the windows has been classified as "Grade 3" in the "Transmission loss grades" of the JHPIS. "Grade 3" is equivalent to "Level 5," the highest level in the assessment levels of the CASBEE's relevant item, which has satisfied the desired value, "Level 4 or over." In addition, "Grade 3" in the "Transmission loss grades" meets 25db or more of sound transmission loss, measured by one third octave band analysis [5].

The entrance door of the home is equipped with two locks. The three double-sliding windows which face the veranda are equipped with shutters. These installations are equivalent to "effective measures to prevent intrusion" described in the section of the "Precautions against crimes" of CASBEE. On the other hand, only basic measures to prevent intrusion have been taken for the outward opening windows in the bedroom and kitchen. As a result, this home's anti-crime performance has been assessed at "Level 3" in the assessment levels of the relevant CASBEE's section.

As demonstrated in the bottom right-hand photo of **Figure 3**, "shutters" have been installed on the large windows which face south. The installment of these shutters have been aimed at protecting the glass against impacts, including fire, hurricane, and flying objects [1]. This installment can also reduce the risk of being damaged by the adverse impacts of climate change, such as typhoons [1], which are projected to increasingly become more severe.

- Equipment for harnessing natural energy

As demonstrated in **Figure 3**, the greater part of the single-pitch roof is covered with solar panels. The total power generation capacity of the 49 solar panels amounts to 11.4 kW. In the 1 year immediately after moving-in, this solar system generated the electricity of 15,911 kWh [1]. In the same period, the energy consumed in this house has been 3085 kWh [1]. Thus, as shown in **Figure 4**, the energy self-sufficiency reached a surprising 516%, which substantially exceeded the desired value, 100% [1].

The solar-generated electricity is preferentially used in this home and the surplus electricity is transmitted to the power grid. The electricity transmitted to the power grid is used by the neighboring electricity consumers including surrounding households.

In addition, equipment for harnessing natural energy secures alternative energy sources and increases resilience, or passive survivability in crises. Therefore, it is recognized as an "adaptation" measure as well as "mitigation" [6–8].

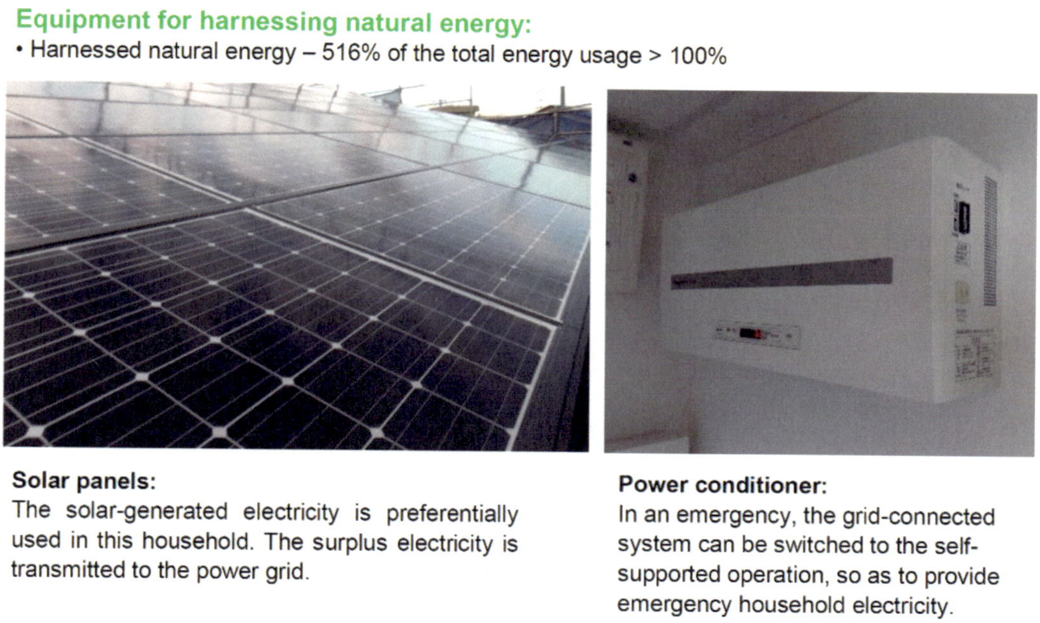

Figure 4. Equipment for harnessing natural energy.

- Framework

This home has been designed so that the "resistance to earthquakes" of the framework meets "Grade 3" in the Seismic resistance grades of JHPIS (**Figure 5**), which surpasses the required level, Grade 2. "Grade 3," the highest grade, means that the building can withstand 1.5 times the strength of an earthquake stipulated in the Building Standards Act of Japan, as an earthquake that occurs very rarely, that is, once every several 100 years [5].

The "durability" of the framework meets the desired value, that is, "Grade 3" in the Deterioration resistance grades of JHPIS [1]. "Grade 3," the highest grade, requires measures to extend the period of time between the construction and the first large-scale renovation up to three generations (about 75–90 years) or more, under ordinarily assumed conditions and maintenance [5].

More than half of the "material" used in this wooden house's framework has been produced from domestic sustainable forests. More specific, we have used wood from Fukushima Prefecture, Tohoku Region. In addition, the legality and sustainability of the wood are certified by a Japanese forestry organization, the *Mokuzai-Hyoji-Suishin-Kyogikai*, the English name of which is the Forest-products Identification Promotion Conference (FIPC). In the bottom left-hand photo of **Figure 5**, the mark of "FIPC" can be seen on the label put on the wood. These considerations meet the criterion of "Level 5," the highest level, of the CASBEE's "Materials useful for resource saving and waste prevention" [4].

- Thermal insulation

Rigid polyurethane foam and extruded polystyrene form are used as thermal insulation materials of this home. Boards of rigid polyurethane foam, one of the most efficient high performance insulation material with minimal occupation of space [9, 10], are placed in the walls facing outside as well as on the ceiling of the second floor (**Figure 5**, right). Boards of extruded

Figure 5. Framework and thermal insulation.

polystyrene foam, features of which includes greater insulating power and higher resistance to compression and moisture [9, 10], are installed on the base and earth slab concrete.

The installation of the above thermal insulation boards enables the home's insulation performance to achieve the desired value, the highest grade in the "Energy-saving action grades" of JHPIS. To be concrete, the heat loss coefficient (Q) of this house is "$Q = 1.90\ W/(m^2 * K)$." This value is substantially better than the target value of heat loss coefficient of the classified area where Tokyo is included, namely "$Q = 2.70\ W/(m^2 * K)$."

Figure 6 illustrates three material elements, namely "interior," "appliances," and "lighting fixtures".

Figure 6. Interior, appliances, and lighting fixtures.

- Interior

Interior, including floors, walls, and ceilings, requires "measures against formaldehyde" as a significant variable. Formaldehyde, a colorless, strong-smelling gas, is used in making building materials and many household products; exposure to formaldehyde can cause adverse health effects, such as irritation of the skin, eyes, nose, and throat [11]. We have set the desired value of "measures against formaldehyde" at "Level 5" of the "Countermeasures against chemical contaminants" of CASBEE. "Level 5," which is equivalent to "Grade 3" in the relevant item of the JHPIS, requires that formaldehyde emissions from interior finish and base materials are "extremely low" [4, 5]. Meanwhile, all of the interior finish and base materials used in this home have been "F-four-star" certified products by the Japanese Industrial Standards or Japanese Agricultural Standards. Overall utilization of "F-four-star" certified products has satisfied the desired value [1].

Three kinds of "materials" which promote resource saving or waste prevention have been used for the interior finishing and sheathing materials of this home. The product used for wall sheathing and ceiling sheathing is plasterboard made of desulfurized plaster. The tiles laid on the entrance floor are made of waste silica sand and waste clay from ceramic industry. Interior fixtures such as the sliding doors and door frames are made of recycled medium density fiberboard. On the other hand, we could not use recycled, renewable and recyclable materials for the wall and ceiling finishing and floor sheathing. As a result, the variable "materials" of interior has been evaluated at "Level 3" in the assessment levels of the relevant item of CASBEE, which has not reached the desired value.

- Appliances

Home appliances are necessary to be energy-saving devices. We have identified the variable of such appliances as the "Energy-saving standard achievement rate" and set their desired value at "100% or more" in principle. Instead of "100% or more," "three or more stars" has been identified as its desired value for several kinds of appliances including air conditioners, refrigerators and televisions.

All of appliances used in this home are energy efficient. For example, the energy-saving standard achievement rates of the refrigerator, television, and air conditioner are "five stars," "five stars," and "four stars," respectively. These rating levels have satisfied the desired value, "three or more stars."

- Lighting fixtures

All of the "lighting fixtures" of this house use "LED" lights, which meets the desired value.

In addition, the lighting fixtures in the four living spaces, namely the living room, bedroom, study and multipurpose room, are equipped with dimmer device. By operating dimmer remote control, the occupants can adjust the brightness of the light, as need demand.

- Water heater

We have selected an "electric heat pump water heater" as hot water supply equipment. Heat pump water heaters take the heat from surrounding air and transfer it to water in an enclosed tank. Heat pump water heaters use electricity to move heat from one place to another, instead of generating heat directly [12, 13]. Therefore, they can be much more energy efficient than conventional electric resistance water heaters.

The utilization of an electric heat pump water heater meets the desired value, "Level 5" in the assessment levels of the CASBEE's relevant item. In addition, the heat pump water heater used in this home (**Figure 7**) can produce three times as much heat energy as consumed electric energy.

- Piping

"Piping," including drainage pipes, water pipes and gas pipes, needs "measures for maintenance" as an important variable toward a long service life. The "measures for maintenance" of this home has achieved the highest level in the Maintenance grades of the JHPIS, Grade 3 (**Figure 8**, left). The Grade 3 requires "creating openings for cleaning and inspection," in addition to the basic provisions such as "not burying piping under concrete" [5].

"Method of water and hot water piping," the other variable of piping, requires "header and pipe-in-pipe system" as its desired value. Recently, the header and pipe-in-pipe system has

Figure 7. Water heater.

Figure 8. Piping.

been becoming widespread in Japan, due to its several advantages, including easy maintenance, easy piping work, and energy efficiency. Consequently, we have naturally adopted this piping method and achieved the desired value (**Figure 8**, right).

- Water-using equipment

"Water-using equipment," including toilet bowls, faucets, and shower heads, requires "water-saving functions" as its key variable. The water-using equipment of this home includes three types of water-saving functions: water-saving type toilets, kitchen thermostat type water faucet, and bathroom thermostat type water faucet plus water-saving shower head equipped with hand-operated water-shutoff mechanism (**Figure 9**, left). The inclusion of the three functions has led to "Level 5" in the five levels of the "Water-saving systems" of CASBEE [4].

- Bathtub

As illustrated in the right of **Figure 9**, the bathtub used in this home is "insulated," which meets the desired value. This type of bathtub sets conforms to the "High thermal insulation bathtub" of Japan Industrial Standards. This insulated bathtub and lid can control temperature drop of the warm water in the bathtub within 2.5° as long as 4 hours [14].

Figure 10 illustrates four spatial elements, namely "bathroom," "toilet," "hallway," and "stairs."

- Bathroom

When designing this home, we have selected a relatively large prefabricated bath unit with a handrail. In addition, recently modular bathrooms are widely used in Japan. The internal width of this bath unit is 160 cm, which has met the criterion, namely 130 cm or more. The internal floor

http://dx.doi.org/10.5772/intechopen.71323

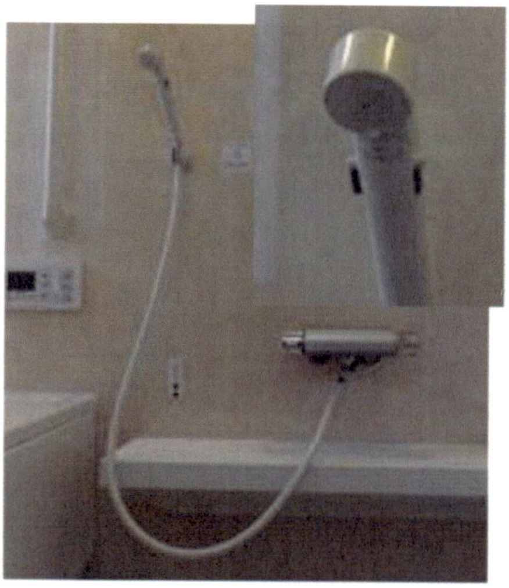

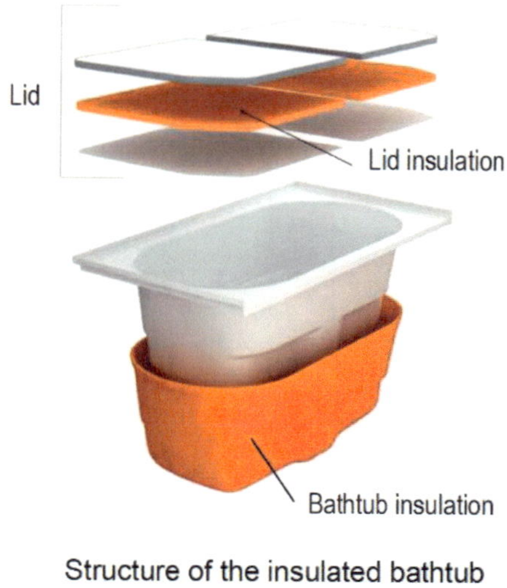

Figure 9. Water-using equipment and bathtub [14].

space of this bath unit is 3.2 m^2, which also has satisfied the criterion, 2 m^2 or more. Meanwhile, the installed handrail can help users to go in and out of the bathtub.

- Toilet

The toilet on the first floor has a larger area than usual Japanese toilet areas. The internal length of the space is 169 cm, which has met the criterion, namely 130 cm or more. The spacing from the front rim of the toilet bowl to the wall is 98 cm, which has sufficiently surpassed the target figure, 50 cm.

As shown in the bottom left-hand photo of **Figure 10**, "handrails" are installed beside the toilet bowl, so as to help users to sit and stand. The vertical handrail is a common cylindrical bar; the horizontal handrail doubles as a shelf.

- Hallway

We have designed the hallway, which connects the entrance and the doors for the living and dining room and the toilet, considering future possibilities of wheelchair usage. The "width" of this hallway is 104 cm, which has satisfied the desired value, 78 cm or more.

- Stairs

We have designed the stairs in this home, so as to meet higher safety requirements. The three numerical values relating to "grade of steepness" of the stairs, namely (1) rise/run, (2) rise * 2 + run, and (3) run, are "18/25," "610," and "250 mm," respectively. These figures have satisfied the following criteria: (1) rise/run ≤22/21, (2) 550 mm ≤ (rise * 2 + run) ≤ 650 mm, and

Bathroom:

- Width – 160 cm > 130 cm [OK]
- Floor space – 3.2 m^2 > 2 m^2 [OK]
- Handrail helps users go in out of the bathtub – Installed [OK]

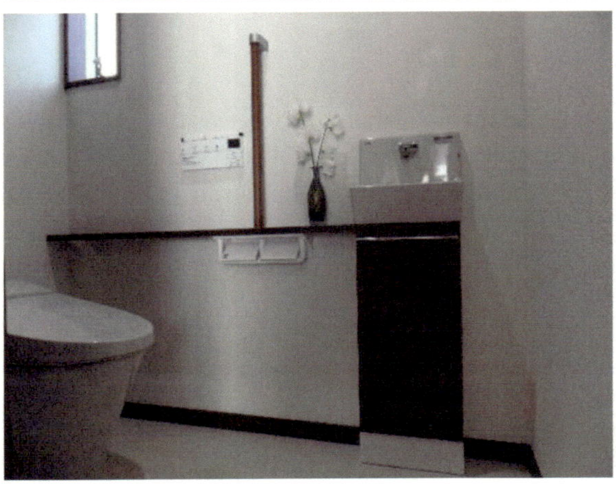

Toilet:

- Length – 169 cm > 130 cm [OK]
- Spacing from the front rim of the toilet bowl to the wall – 98 cm > 50 cm [OK]
- Handrails help users sit and stand – Installed [OK]

Hallway:

- Width – 104 cm > 78 cm [OK]

Stairs:

- Grade of steepness:
  Rise/Run – 18/25 < 22/21 [OK]
  (Rise * 2 + run) – 610 mm < 650 mm [OK]
  Run – 250 mm > 195 mm [OK]
- Handrail – Installed [OK]

Figure 10. Bathroom, toilet, hallway, and stairs.

(3) run ≥195 mm. Accordingly, the "grade of steepness" of the stairs fulfills the desired value, "Grade 3" of the JHPIS's relevant item. Meanwhile, a "handrail" has been "installed" on one side, which has also met the desire value.

- Doorways

Figure 11 demonstrates "doorways," focusing on "difference in level" and "width." All of the doorways in this home have no difference in level. The width of the doorways on the first floor except for the bathroom is 75 cm and that for the bathroom is 67 cm, both of which have met the desired values, 75 cm and 60 cm.

In addition, at the entrance door, traditionally stepped thresholds help with weather protection. Level access requires consideration of alternative solutions to maintain appropriate protection from the wet weather. In this case, a grated drain and a larger cover over the porch play a role of limiting the quantity of water around the entrance door area.

- Main access route to the entrance and slope

In the case of this home, "main access route to the entrance" is the path from the southern frontal road to the entrance porch, as shown in the upper left photo of **Figure 12**. In order to achieve stepless approach, the area that connects the frontal road and the porch has a sloping "surface." Meanwhile, the "width" of this access route is 140 cm, which has easily met the desired value, 90 cm or more.

The level difference between the frontal road and the porch is 45 cm; the horizontal distance of the slope is 410 cm. Consequently, the "grade of steepness" of the "slope" is approximately 1/9, which reaches the desired value, 1/8 or less. Moreover, a "handrail" has been "installed" on one side, which has also satisfied the desire value.

- Outdoor facilities (Fence, etc.)

When designing the exterior of this home, we have strived to make it open and ecological. The approach area on the west side of this lot is not enclosed by any fences and door gates. The garden area, which covers two-third of the exterior area, is not also fenced. However, we have surrounded the veranda with a fence, in order to secure privacy and prevent crimes. This section evaluates this home's "outdoor facilities," including the fence, veranda deck, porch roof, and approach floor, from the perspective of the three variables: "form," "appearance," and "materials."

"Form," the desired value of which is set at "not blocking sightlines," is significant for fences and barriers, in particular. The vertical-latticed and 180 cm-high fence screens the home and the washing hung in the veranda from the public view. On the other hand, any suspicious person hiding behind the fence can be seen through the openings of the vertical lattice. Meanwhile, the height of the fence from the veranda deck level is 130 cm; therefore, a person standing on the veranda deck can communicate with neighboring residents without being hindered, as shown in the right photos of **Figure 12**. In this way, this fence's form does not block sightlines and contribute to "safety" and "mutual help" in this area, through preventing crime and allowing face-to-face communication.

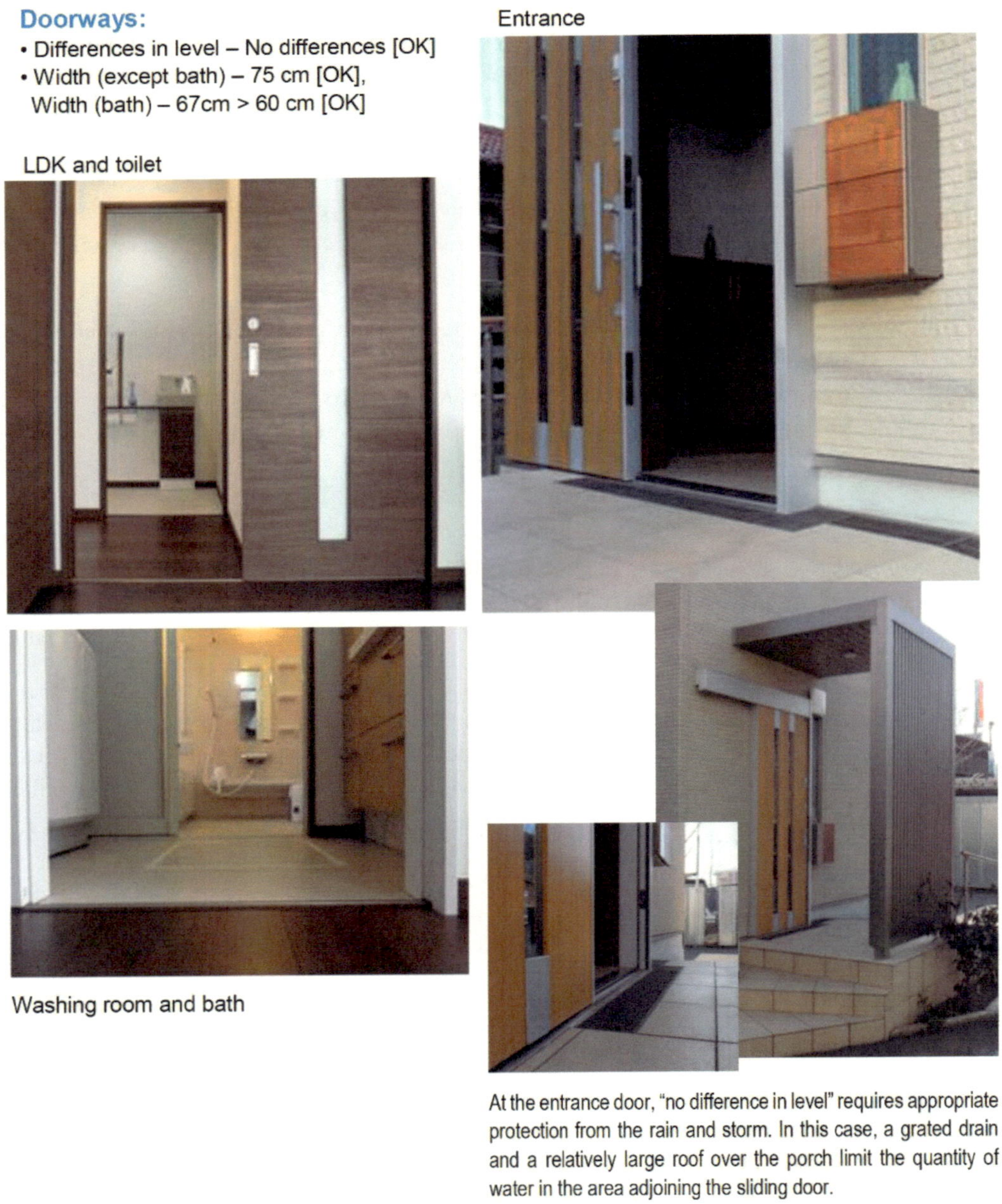

Figure 11. Doorways.

"Appearance" of outdoor facilities, including color and shape, requires "consideration for the landscape." The color of the fence and the porch's roof and screen, which is called "shining gray," is the same as that of the house's window frames and shutters. The tiles of the approach floor have a similar color to that on the outer walls, namely light beige. We expect that the above color coordination creates sense of unity between the house and outdoor facilities and moreover brings stability to the landscape.

http://dx.doi.org/10.5772/intechopen.71323

Figure 12. Main access route to the entrance/slope and outdoor facilities.

Furthermore, when selecting outdoor home products, we have also paid attention to their "materials" and "sustainable use of natural resources." The floor tile of the approach is a recycled product made of waste silica sand and waste clay from ceramic industry. The fence and the porch roof and screen are made of recyclable and long-life aluminum. The veranda deck is made of artificial materials produced from wood powder and the thermoplastic resin. These considerations have easily enabled the assessment of the "materials" to reach its desired value, "Level 5" of the relevant item of CASBEE.

Figure 13. Equipment for rainwater use.

- Equipment for rainwater use

"Equipment for rainwater use," in this case, a 200-L rainwater tank, meets the desired value. As demonstrated in **Figure 13**, the rainwater in this tank is used for watering the garden. Using rainwater can reduce the quantity of water supply. At the same time, storing rainwater is also considered one of "adaptation" measures because it leads to securing emergency water supply [6–8].

4. Performance evaluation

4.1. Quantitative evaluation

4.1.1. Evaluation by CASBEE for detached houses

After the home started to be used, we had its sustainability evaluated by the "CASBEE for Detached Houses (New Construction, 2010 edition)."

"CASBEE for Detached Houses" evaluates the comprehensive environmental performance of detached houses based on the following six categories: Comfortable, healthy and safe indoor environment (Q1), Durability for long-term use (Q2), Consideration for the townscape and ecosystem (Q3), Energy and water conservation (L1), Conservation of resources and reduction of waste (L2), and Consideration for the global, local and surrounding environment (L3). Dividing "Q" by "L," CASBEE represents building environmental performance with a single score of Built Environment Efficiency (BEE). According to the BEE value, detached houses are graded into five ranks: S (Excellent, five stars), A (Very good, four stars), B+ (Good, three stars), B- (Fairly poor, two stars), and C (Poor, one star) [15].

The evaluation results show that the "Q" and "L" have been "88" and "13," respectively. Consequently, as demonstrated in the left of **Figure 14**, the BEE score has reached 6.7, which has rated and certified the home as the highest "S," or "five stars" with ease. All scores of the six categories are high, namely Q1 = 4.4, Q2 = 4.6, Q3 = 4.6, L1 = 4.7, L2 = 4.0, and L3 = 4.6, as shown in the right of **Figure 14** [16]. These values prove that this house excels in both environmental quality and environmental load reduction.

4.1.2. Life-cycle-carbon-minus (LCCM) house certification

After the CASBEE evaluation, this home was certified as a life-cycle-carbon-minus (LCCM) house. The LCCM house is defined as a "house with a negative CO_2 emission in its total life cycle, including construction, utilization and demolition" [17]. There are several key factors in achieving an LCCM house, that is, a long service life, least amount of CO_2 emissions for constructing, using and demolishing the house, and the utilization of renewable energy, such as solar power generation [17]. LCCM House Certification requires that the house is rated "S" or "A" of CASBEE rating, as its precondition. The Certification is classified into "five stars" and "four stars;" only houses with "five stars" meet the above definition.

The Life Cycle CO_2 of this house has been considerably less than zero, due to its higher energy-saving performance and larger solar power generation capacity. As shown in the left of **Figure 15**, it has been "minus 35%," as compared with 100% of common houses' Life Cycle CO_2, or the reference value [18]. As a result, the LCCM House Certification has also been classified into "five stars" with ease. In addition, as of July 2016, the number of LCCM certified houses with "five stars" is 17 nationwide, including this house [19]. In Tokyo Metropolis, this house has been the first one.

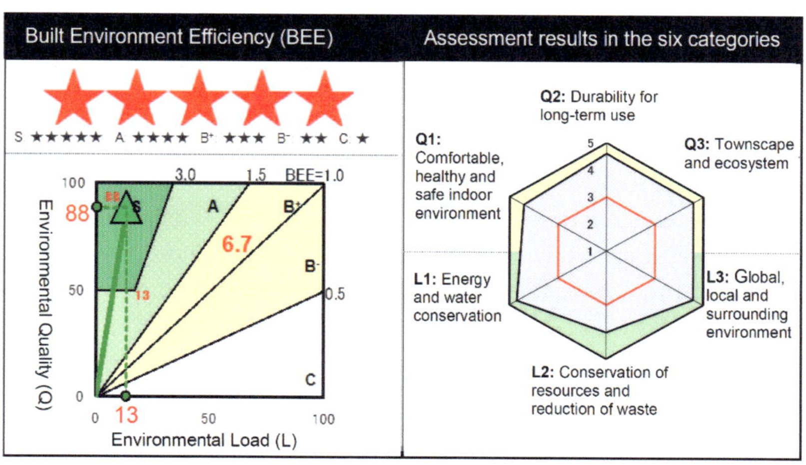

Figure 14. CASBEE assessment results of this home [16].

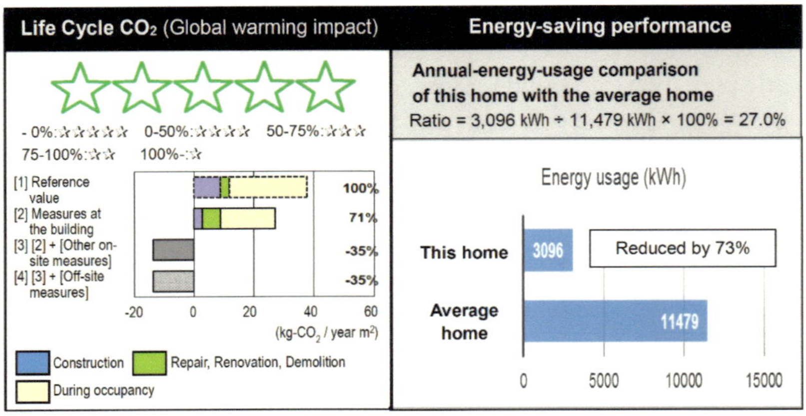

Figure 15. Life cycle CO_2 analysis [18] and energy-saving performance of this house [1].

4.1.3. Energy-saving performance

We have assessed the energy-saving performance of this house, by comparing its annual energy consumption with that of the average house. The total energy consumption of this house per year was 3085 kWh [1]. On the other hand, the average annual energy consumption of the same two-person-household detached houses amounts to 41,325 MJ [20], which is equal to 11,479 kWh. This home's annual usage, 3085 kWh, is equivalent to 27.0% of 11,479 kWh. Accordingly, as illustrated in the right of **Figure 15**, this home has reduced energy usage by over 70%, as compared to the average home of the same condition [1].

4.1.4. Water-saving performance

We have also checked the water usage. The monthly average water consumption of this home was 12.8 m^3, whereas that of the same two-person-household homes was 16.2 m^3 [21]. Accordingly, this home has saved water consumption by 21%, in comparison to the ordinary home of the same condition.

4.2. Reactions of the occupants and visitors

After the moving-in, the occupants, the owner couple, expressed various impressions on this home as well as differences between living in it and in their previous home. Meanwhile, the visitors gave their reactions to this home and its various parts. Choosing several points from such impressions and reactions, this section describes these points, in line with related elements and their variables.

4.2.1. Thermal insulation performance

"Higher thermal insulation performance" has contributed to more stable indoor temperature and less demand for heating and air-conditioning. These favorable effects have naturally led to comfort and energy conservation, as we expected.

Furthermore, another remarkable change which occurred to the occupants was a significant improvement in allergy symptoms of the owner wife. While living in the previous house, she suffered from cedar pollen allergy every spring. In the spring immediately after moving-in, her symptoms sharply changed for the better. Also in the following springs her improved condition has continued. The mechanism which has brought this symptomatic relief is uncertain; however, there is a strong possibility that higher thermal insulation performance of this home has influenced it. The previous home's windows, door, and walls facing outside had poor thermal insulation performance; as a result, water condensation and mold frequently appeared. On the other hand, water condensation and mold have rarely been observed in this home since the thermal performance of the building envelope is superior as a whole.

In addition, there have been many studies which indicate a causal relationship between upgrading of the thermal insulation performance and decrease in the occupants' prevalence rates of various diseases including allergies [22, 23]. Accordingly, this case is considered as one example of supporting such a causal relationship.

4.2.2. Areas relating to water use and hot water supply

Originally, "placing areas relating to water use and hot water supply closer" has been intended to reduce materials for piping and heat loss from hot water piping. After starting to use this home, the occupants have noticed that this consideration also leads to comfort and convenience in daily life. First, placing such areas closer can reduce time until hot water comes out. Moreover, in this case, the floor planning, where the kitchen, laundry space, and bath are arranged closer together, has naturally shortened working route and improved efficiency of household tasks.

4.2.3. Main access route to the entrance

When designing the main access route to the entrance as a slope, we mainly aimed to prepare for the occupants' future. However, this sloped access route has already gained approval beyond our expectation. As a woman visited this home, she pointed out that this slope is favorable for wheeling a stroller or baby carriage, in addition to using a wheelchair. Moreover, when the owner wife experienced difficulty in walking due to a leg injury, she had a real feeling that using the slope is much easier than going up or down the stairs. These remarks show that a step-free approach enables a variety of people to have easier access.

4.2.4. Doorways

In addition to the step-free approach, the level and slightly wider doorways have also been received favorably on various occasions. The occupants have never stumbled over doorways in this home, unlike in their previous home. Furthermore, they often feel comfortable and unimpeded movement between spaces, for instance, when moving furniture or baggage, and vacuuming or cleaning the floors. These experiences show that flat and wider doorways can bring safety and convenience as well as easy access for people with mobility difficulties.

4.2.5. Protection of glass against impacts

We have installed “shutters” on the large windows facing south, in order to protect glass, the most vulnerable part of housing exterior. This installment is expected to reduce the risk of being damaged by current and future impacts such as fire and serious extreme weather events. Moreover, this consideration has offered the occupants other benefits, that is to say, it has given them a sense of security and enhanced thermal insulation performance. In cases where a typhoon is approaching or a tornado warning is issued, closing shutters brings a feeling of safety. Installing shutters is also useful to prevent intrusions and increase the home’s anti-crime performance. Furthermore, the closed shutters increase thermal insulation performance, which is effective especially on winter nights, by adding an air layer between the windows and shutters.

4.3. Other evaluation: sustainable housing award

This home has won the Japan Wooden Houses & Industry Association President Award, in the Sixth Sustainable Housing Awards [24]. The Sustainable Housing Awards is a housing design competition which is held every 2 years under the auspices of the Institute for Building Environment and Energy Conservation [25] of the Ministry of Land, Infrastructure, Transport and Tourism. What are publicly commended at the competition are excellent detached houses that combine a comfortable living environment while lowering the burden on the natural environment [25].

In the selection process of the Sixth Sustainable Housing Awards, the screening committee, which was composed of 11 experts including 7 university professors, examined 46 entries [26];

Figure 16. Certificate of the sustainable housing award, CASBEE assessment, and LCCM house.

as a result, this home has received that award. The award review states that the main attractive points of this home have been its strikingly high energy-saving efficiency and energy self-sufficiency [27]. Furthermore, this home has been highly assessed, due to the certification as a Life-cycle-carbon-minus (LCCM) house with five stars as well as the extremely high value of Built Environment Efficiency (BEE) in the "CASBEE for Detached Houses" evaluation (**Figure 16**) [27].

Author details

Kazutoshi Fujihira

Address all correspondence to: fujihira@kankyogaku.com

Institute of Environmentology, Tokyo, Japan

References

[1] Fujihira K. System control for sustainability: Application to building design. In: Thomas C, editor. Complex Systems, Sustainability and Innovation. Rijeka, Croatia: InTech; 2016. p. 37-55. DOI: http://dx.doi.org/10.5772/65875

[2] Center for Better Living, Association for Housing Performance Evaluation and Indication. Long-Life Quality Housing Certification Manual (in Japanese) [Internet]. 2010. Available from: http://www.hyoukakyoukai.or.jp/kikan/pdf/chouki_manual.pdf [Accessed: 2017-03-17]

[3] Ministry of Land, Infrastructure, Transport and Tourism. Japan Housing Performance Indication Standards Evaluation Method Criteria (in Japanese) [Internet]. Available from: http://www.mlit.go.jp/common/000052960.pdf [Accessed: 2017-04-11]

[4] Japan Green Build Council, Japan Sustainable Building Consortium. CASBEE for Detached Houses (New Construction) – Technical Manual, Edition (in Japanese). Tokyo, Japan: Institute for Building Environment and Energy Conservation; 2010. p. 178

[5] Ministry of Land, Infrastructure, Transport and Tourism. Japan Housing Performance Indication Standards (in Japanese) [Internet]. 2001. Available from: http://www.mlit.go.jp/jutakukentiku/house/torikumi/hinkaku/061001hyouji.pdf [Accessed: 2016-06-10]

[6] IPCC. Climate Change 2007: Synthesis Report. Contribution of Working Groups I, II and III to the Fourth Assessment Report of the Intergovernmental Panel on Climate Change. IPCC: Geneva, Switzerland; 2007. p. 104

[7] Wilson A. Passive Survivability: A New Design Criterion for Buildings [Internet]. 2006. Available from: https://www2.buildinggreen.com/article/passive-survivability-new-design-criterion-buildings [Accessed: 2017-04-10]

[8] Ward A, Wilson A. Design for Adaptation: Living in a Climate-Changing World [Internet]. 2009. Available from: https://www.buildinggreen.com/feature/design-adaptation-living-climate-changing-world [Accessed: 2017-04-10]

[9] XCO2 conisbee. Insulation for Environmental Sustainability: A Guide [Internet]. Available from: http://kingspaninsulation.co.uk/getattachment/dcbc3aca-1f77-411e-ada5-39c186c4cf62/White-Paper--Insulation-for-Environmental-Sustaina.aspx [Accessed: 2017-04-10]

[10] Dannetsujyutaku.com. Features of Various Kinds of Thermal Insulation (in Japanese) [Internet]. Available from: http://dannetsujyutaku.com/basic/knowledge/spec [Accessed: 2017-02-25]

[11] United States Environmental Protection Agency. Facts about Formaldehyde [Internet]. Available from: https://www.epa.gov/formaldehyde/facts-about-formaldehyde [Accessed: 2017-03-17]

[12] Office of Energy Efficiency & Renewable Energy. Heat Pump Water Heaters [Internet]. Available from: https://energy.gov/energysaver/heat-pump-water-heaters [Accessed: 2017-02-24]

[13] Energy Star. How it Works—Heat Pump Water Heaters [Internet]. Available from: https://www.energystar.gov/products/water_heaters/high_efficiency_electric_storage_water_heaters/how_it_works [Accessed: 2017-02-24]

[14] LIXIL Corporation. System Bathroom Kireiyu (in Japanese). Tokyo, Japan: LIXIL Corporation; 2012. p. 170

[15] Murakami S, Iwamura K, Cole R. CASBEE: A Decade of Development and Application of an Environmental Assessment System for the Built Environment. Tokyo, Japan: Institute for Building Environment and Energy Conservation; 2014. p. 297

[16] Institute for Building Environment and Energy Conservation. CASBEE for Detached Houses (New Construction) Evaluation Results: Inagi EcoHouse (in Japanese) [Internet]. Available from: http://www.ibec.or.jp/CASBEE/DH_certification/kohyo/IBEC-L0040H.pdf [Accessed: 2016-06-10]

[17] Building Research Institute, Japan. What Is an LCCM House? [Internet]. Available from: http://www.kenken.go.jp/english/contents/lccm/index.html [Accessed: 2017-03-01]

[18] Institute for Building Environment and Energy Conservation. LCCM Assessment Results: Inagi EcoHouse (in Japanese) [Internet]. Available from: http://ibec.or.jp/rating/list/pdf/IBEC-L0040H.pdf [Accessed: 2017-03-01]

[19] Institute for Building Environment and Energy Conservation. List of the LCCM Certified Houses (in Japanese) [Internet]. Available from: http://www.ibec.or.jp/rating/lccm-list.html [Accessed: 2017-03-01]

[20] Mitsubishi Research Institute, Research Division of Environment and Energy. Survey of Energy Consumption in Fiscal 2012: Fact-Finding Survey on Energy Consumption in the

Consumer Sector (in Japanese) [Internet]. 2013. Available from: http://www.meti.go.jp/meti_lib/report/2013fy/E003078.pdf [Accessed: 2016-06-10]

[21] Bureau of Waterworks Tokyo Metropolitan Government. What we would Like to Know about Water Supply: Average Water Usage of Households (in Japanese) [Internet]. Available from: https://www.waterworks.metro.tokyo.jp/faq/qa-14.html [Accessed: 2017-02-15]

[22] Howden-Chapman P, et al. Effect of insulating existing houses on health inequality: Cluster randomised study in the community. BMJ. 2007;**334**:460-464. DOI: http://dx.doi.org/10.1136/bmj.39070.573032.80

[23] Ikaga T, Eguchi R, Murakami S, Iwamae A, Hoshi T, Mizuishi T, Kawakubo S, Okumura K. Evaluation of investment in residential thermal insulation considering non-energy benefits delivered by health (in Japanese). Journal of Environmental Engineering (Transactions of AIJ). 2011;**76**:735-740. DOI: http://doi.org/10.3130/aije.76.735

[24] Institute for Building Environment and Energy Conservation. Japan Wooden Houses & Industry Association President Award in the Sixth Sustainable Housing Awards: Inagi EcoHouse (in Japanese) [Internet]. Available from: http://www.ibec.or.jp/hyosyou/housing/6th/pdf/04.pdf [Accessed: 2017-03-01]

[25] Institute for Building Environment and Energy Conservation. Sustainable Housing Awards (in Japanese) [Internet]. Available from: http://ibec.or.jp/sustainable/housing/index.html [Accessed: 2017-03-01]

[26] Institute for Building Environment and Energy Conservation. Concerning "the sixth sustainable housing awards" (in Japanese). IBEC. 2015;**36**(1):2-3

[27] Shiraishi Y. Japan Wooden Houses & Industry Association President Award: Inagi EcoHouse (in Japanese). IBEC. 2015;**36**(1):10-11

Chapter 6

Discussion and Conclusion: Effectiveness, Characteristics and Future Prospects of the Methodology

Kazutoshi Fujihira

Additional information is available at the end of the chapter

http://dx.doi.org/10.5772/intechopen.71321

Abstract

Following the control system for promoting sustainable home design, we have designed a home and built it. The quantitative evaluations and reactions of the occupants and visitors on the home indicate that if system users closely follow the methodology, they can comprehensively achieve sustainable homes, which have high environmental performance. Meanwhile, the results of the study have suggested that this methodology has several characteristics, besides comprehensiveness. First, the diagram of the control system itself is useful because it concisely explains the whole picture of the sustainable design processes on both new and existing homes. Second, the "sustainable design guidelines" and "sustainability checklist" are user-friendly since the material and spatial elements are equivalent to real parts of homes. Moreover, the "element – variable – desired value" structure in the guidelines and checklist is superior in "adaptability to regional differences" and "flexibility toward changes over time." We expect that this methodology is widely used, in coordination with the existing methods for sustainable housing. Furthermore, it can be theoretically applied to other categories of human activities, which are regarded as the complex of material and spatial elements.

Keywords: design process, visualization, "element – variable – desired value" structure, comprehensiveness, user-friendliness, adaptability, flexibility

1. Introduction

The last chapter has illustrated a case study in which a home has been designed and constructed, based on the methodology of applying control science to sustainable housing design. Reviewing the methodology and case study, this chapter explores the effectiveness, characteristics, and future prospects of the methodology. The discussion section also examines the capability of the methodology for the two significant changes, namely climate change and aging population, which have been presented in Chapter 1. Furthermore, the same section discusses how this methodology deals with the remaining issues in the existing Japanese systems, which have been raised in Chapter 2. After that, the final section summarizes the conclusions of the studies.

2. Discussion

This section discusses the results of this study from the following five perspectives: (1) effects of the methodology for promoting sustainable home design, (2) characteristics of the methodology, (3) coordination with the existing systems, (4) applicability of the methodology, and (5) future research.

2.1. Effects of the methodology for promoting sustainable home design

As illustrated in the case study, we have designed a detached house, following the methodology for promoting sustainable home design. That is to say, we have designed the house's parts or elements, so that the elements' variables satisfy their desired values as much as possible. After the house began to be used, we have obtained objective evaluations of the home's environmental and sustainability performance as well as various comments on the home from the occupants and visitors.

First, all of the objective and quantitative evaluations have shown that the home has considerably high environmental performance. According to CASBEE for detached houses, a comprehensive assessment system, the home has readily been ranked in the highest "S," with an extremely high score of built environment efficiency (BEE). Subsequently, the house has been classified into the highest "five star" by life cycle carbon minus (LCCM) certification, due to its higher energy saving performance and larger solar energy generation capacity. Meanwhile, the energy usage comparison with the average home has shown that the total energy usage of this home is equal to about 27% of the average home under the same conditions. The water usage comparison with the average home has also proved the higher water saving performance of this home. In addition, such high environmental performance has been highly evaluated when this house received the prize at the Sustainable Housing Awards.

The reactions of the occupants and visitors suggest that following the methodology has produced various favorable effects. On the other hand, any unfavorable side effects of utilizing the methodology have not been observed thus far. "Higher thermal insulation performance" has contributed to improve the occupant's allergy symptoms, as well as improving comfort and reducing demand for heating and air-conditioning. "Placing areas relating to water use and hot water supply" has brought about reduction in time until hot water comes out, in addition to the reduction in materials for piping and energy for hot water supply. "Taking accessible and universal design into spatial elements, such as doorways and main access route to the entrance," has already brought about present safety and comfort, above and beyond preparing the occupants for the future. "Protection of glass against impacts with shutters" has also contributed to prevent crimes and increase thermal performance, in addition to reduce the risk of being damaged by impacts such as fire and serious extreme weather events.

In this way, we have evaluated the home, which has been designed closely based on the methodology, from both the quantitative analyses and reactions of occupants and visitors. First, all of the quantitative analyses have shown that the home has significantly high environmental performance. The reactions of the occupants and visitors have also suggested that this home is comprehensively sustainable and comfortable. These assessment results indicate that if system users closely follow the methodology, they can comprehensively realize sustainable homes, which have high environmental performance.

2.2. Characteristics of the methodology

The characteristics of the methodology includes (1) visualization of the whole picture for promoting sustainable home design, (2) user-friendliness, (3) comprehensiveness, (4) adaptability to regional differences, and (5) flexibility toward changes over time.

In addition, *Sustainable Design: A critical Guide*, a book on sustainable building design, mentions four conditions of the ideal method for sustainable building design, that is, "holistic," "flexible," "responsive to local conditions," and "not overly complex to administer" [1]. In this context, "holistic" and "not overly complex to administer" are similar to "comprehensive" and "user-friendly," respectively. Therefore, we consider that this methodology covers all of the above-mentioned four conditions that the ideal method requires.

2.2.1. Visualization of the whole picture for promoting sustainable home design

Figure 5 in Chapter 4 has demonstrated the control system for promoting sustainable home design. This figure basically contains "sustainability," "environmental, social, and economic problems," and "disturbances" as system components. Moreover, "adaptation" to disturbances has been incorporated as a route to sustainability, as well as "solution" or "prevention" of the problems. Utilizing this basic scheme, this figure inclusively shows processes for promoting sustainable design on both new and existing homes with the "sustainable design guidelines" and "sustainability checklist."

We consider that Figure 5 in Chapter 4 concisely explains the whole picture of the sustainable design processes with the guidelines and checklist. Accordingly, we expect that this visualization itself helps people concerned to easily understand that whole picture.

2.2.2. User-friendliness

The "material and spatial elements" in the sustainable design guidelines and sustainability checklist are equal to "actual parts of houses." Thus, the system users can smoothly design, check, evaluate, and inspect the house, by easily comparing the drawings or actual house with the guidelines or checklist [2]. In fact, the design process in the case study has supported the user-friendliness of the guidelines and checklist; the designers of the homebuilder readily accepted them and efficiently made the house's drawings [2]. In addition, "correspondence between the elements and actual parts of houses" is unique to this methodology. On the other hand, the major existing Japanese methods, namely the housing performance indication system (HPIS), long-life quality housing (LQH) certification, and CASBEE for detached houses, do not possess this characteristic, as shown in Chapter 2.

Meanwhile, user-friendliness basically requires such methods to be "not complex" and "not long." As cited before, *Sustainable Design: A critical Guide* has mentioned "not overly complex to administer" as a condition of the ideal method [1]. Similarly, *The Checklist Manifesto: How to Get Things Right*, which impresses the value of checklists for avoiding failures, says that "the checklist cannot be lengthy" [3]. We consider that the "sustainable design guidelines" and "sustainability checklist" meet the above basic requirement. The "guidelines" and "checklist" are relatively simple and compact tables. Each of them fits to two pages of this book, although these functions cover all important elements of homes.

2.2.3. Comprehensiveness

Originally, the "sustainable design guidelines" has been aimed comprehensively at showing the relationships between the standard home and sustainability. Therefore, we expected that following this methodology would lead to achieve comprehensive sustainable homes. The evaluation results of the home in the case study have been obtained as we expected. The CASBEE assessment results, namely the very high BEE score and high scores in all the six categories (Figure 14 in Chapter 5), have supported the comprehensiveness of the methodology as well as the comprehensive sustainability of that home.

2.2.4. Adaptability to regional differences

This methodology, more specific the "element - variable - desired value" structure in the "sustainable design guidelines" and "sustainability checklist," originally has a mechanism of easily adapting to regional differences. As shown in Figure 3 in Chapter 3, examining the relationships between important elements and stability conditions, system designers determine the elements' variables and their desired values. This determination process has a mechanism of reflecting a variety of regional characteristics, including natural, geographical, social, and cultural features [2].

For instance, "resistance to earthquakes," a variable of the framework, reflects a geological feature of Japan, namely "earthquake-prone." This mechanism also enables the system designers to readily vary the guidelines, according to the region's characteristics [2]. For example, if the region is in a strong wind area or snowy area, they can easily adjust the guidelines to the region, by adding "resistance to wind" or "resistance to snow load" as a variable of the framework [2].

Another example of reflecting regional features is "heat insulation" of the bathtub. We have attached importance to this variable, due to a Japanese cultural feature. Reducing heat loss from bath is important in Japan since people frequently take a bath and usually share the same hot water in the bathtub with their family members. On the other hand, in societies without such a lifestyle, it is easy for system designers to simply omit this variable.

http://dx.doi.org/10.5772/intechopen.71321

2.2.5. Flexibility toward changes over time

The "element – variable – desired value" structure in the guidelines and checklist also leads to flexibility toward changes over time. **Figure 1** explains this characteristic, mainly focusing on two major global changes progressing in the twenty-first century, namely "climate change" and "aging population."

First, the course from "climate change" toward the materials or spatial elements passes through "mitigation measures/adaptation measures." This course shows that system designers can take necessary mitigation/adaptation measures against climate change, by adjusting relevant elements, variables, and desired values. For example, when taking a measure of "improving thermal insulation performance," we have added a variable "thermal insulation performance" to two material elements, namely "thermal insulation" and "windows and doors." Similarly, we have easily taken an adaptation measure, by adding "protection of glass against impact" and "with shutters" as a variable and its desired value of the material element, "windows and doors."

In addition, the broken dividing line between "mitigation measures" and "adaptation measures" in the block means that the two types of measures overlap each other. Such overlapping measures include "improving thermal insulation performance," "harnessing natural energy," "utilizing rainwater," and "improving natural ventilation." For instance, "improving thermal insulation performance" contributes to not only saving energy through reducing demand for heating and air-conditioning but also increasing resilience in extreme weather and crises.

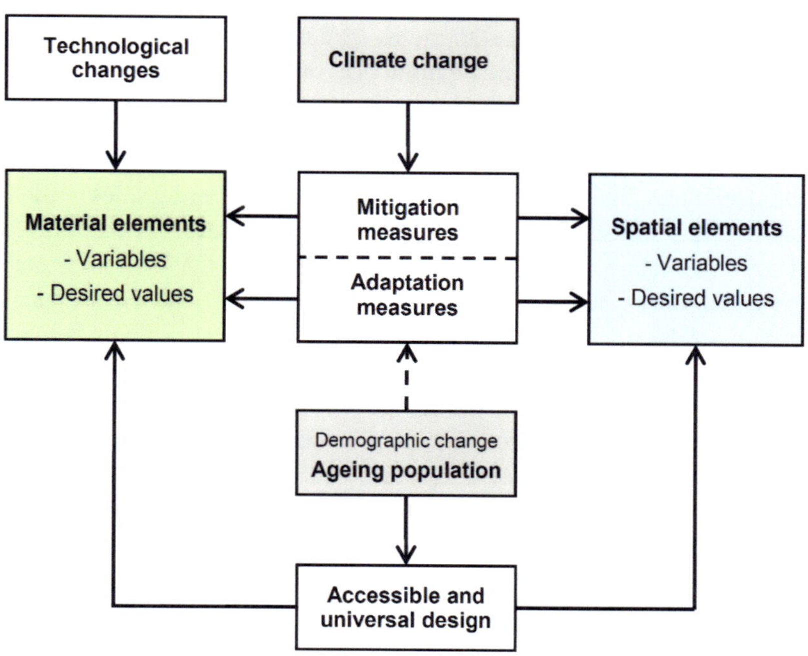

Figure 1. Flexibility of the methodology toward changes over time.

On the other hand, **Figure 1** shows two courses from "aging population" toward the material or spatial elements, that is, the routes through "mitigation measures/adaptation measures" and "accessible and universal design." The broken line that connects "aging population" and "mitigation measures/adaptation measures" means that some of mitigation/adaptation measures benefit elderly people's health. For example, "improving thermal insulation performance" can reduce the number of elderly people's deaths and illnesses resulting from indoor coldness in cold weather. Likewise, "improving thermal insulation performance" and "improving natural ventilation" are expected to decrease heat stroke patients and deaths of the elderly in hot weather.

Meanwhile, the course through "accessible and universal design" represents that system designers can adopt accessible and universal design, by adjusting relevant elements, variables, and desired values. To be concrete, when compiling the guidelines, we first identified spatial elements related to such design, including "specified bedroom," "doorways," "stairs," and "slope." Subsequently, we have added necessary variables to these elements and set their desired values, so as to adopt accessible and universal design.

Furthermore, **Figure 1** contains another change over time, that is, "technological changes." Technological changes, including innovation in technology, are directly related with material elements; therefore, system designers can efficiently take such changes to relevant material elements. For instance, in the latest revision, we have easily taken "LED" as the desired value of "type of light" of "lighting fixtures."

2.3. Coordination with the existing systems

From the beginning, we have intended that this methodology is not used independently but coordinated with the existing Japanese systems that have been shown in Chapter 2. Such coordination is expected to exert favorable influences on the utilization of not only this methodology but also the existing methods.

First, when compiling the "guidelines" and "checklist," we have aimed to provide reliability and incentives to utilize them. The guidelines and checklist use standard grades in the Japan housing performance indication standards (JHPIS) and assessment levels in CASBEE for detached houses, as the desired values of many variables. The JHPIS and CASBEE for detached houses, both of which are national systems, include technical information related to sustainable housing in the descriptions of the standard grades or assessment levels. Accordingly, when referring to the relevant descriptions in the JHPIS and CASBEE for detached houses, system users can obtain reliable related information or technical knowledge about the matters. Meanwhile, following the guidelines enables system users to receive a long-life quality housing (LQH) certification and higher ratings in CASBEE for detached houses. Obtaining a LQH certification leads to preferential treatment from the government, including tax credits; therefore, it can be an incentive for people to use the functions. Moreover, LQH-certified houses and CASBEE higher rated houses have a possibility of competitive superiority in the real estate market. Certified green or sustainable buildings have not yet gained obvious advantages in asset values in the Japanese real estate market, as shown in Chapter 2. However, such competitive superiority is expected to be established sooner or later also in Japan; therefore, it must be another incentive to use the functions.

On the other hand, the utilization of the user-friendly guidelines and checklist has a possibility of promoting the use of the existing public methods. Since the guidelines and checklist refer to CASBEE for detached houses, using these functions naturally leads to the utilization of this trustworthy system but relatively unknown system. Furthermore, when using the "checklist" for inspection or evaluation of existing homes, as a matter of course, "people involved" refer to the "JHPIS (for existing homes)" and "CASBEE for detached houses (existing building)," both of which have hardly been used thus far. The above coordination is expected to produce a synergy effect toward promoting sustainable housing design.

2.4. Applicability of the methodology

2.4.1. Application to other regions and countries

As demonstrated in Section 2.2.4., this methodology, more specific the "sustainable design guidelines" and "sustainability checklist," has a feature of being adaptable to regional differences. Accordingly, it will be not difficult for system designers in another region to adapt these practical functions for that region [2]. That is to say, the system designers can make its regional version, through the examination of the elements and the adaptation of the elements' variables and their desired values to the region's characteristics [2].

As a matter of course, this methodology can be applied to other countries, as well as other regions. When system designers make the guidelines and checklist in other countries besides Japan, they specify variables and their desired values, referring to systems related to buildings and housing used in that country. In such cases, there are two main approaches: (1) specification based on the standards required by building codes and (2) use of criteria shown in voluntary systems related to sustainable housing.

(1) Specification based on the standards required by building codes

If the variables are within the scope of the country's building codes, it is necessary for the system designers to search the building codes for the variables' desired values. Building codes specify the "minimum standards" for constructed objects, in order to protect public health, safety, and general welfare [4]. If system designers consider that the standard value required by the building codes is insufficient for the desired value, they make an addition to the standard value, so as to suit the desired value. On the other hand, if they consider that the standard value is suitable to the desired value, they can use it as it is. In the latter case, the variable and its desired value can be omitted from the guidelines and checklist, for people who naturally conform to the building codes, which have legal force.

In fact, Tables 2 and 4 in Chapter 4 also include variables and their desired values, which have been specified based on the above approach. For example, when determining the desired value of "ratio of total window area to floor area in each living space," a variable of "position and area of windows," we have made an addition to the standard value required by the Building Standards Act of Japan, namely "1/7 (14.3%) or more," and set the desired value at "20% or more." Moreover, "JHPIS 1.1: Grade 2 or over," the desired value of "resistance to earthquakes" of "framework," has been originally determined, based on the standard required by the building code. To be concrete, "Grade 2" in this desired value means 1.25 times the strength of an earthquake stipulated in the Building Standards Act of Japan. In addition, "resistance to winds" and "resistance to snow load" are possible variables of "framework" that have been omitted from the tables, because the standards for these two required by the building code have been considered to be suitable to their desired values.

(2) Use of criteria shown in voluntary systems related to sustainable housing

The second approach is to utilize criteria shown in voluntary systems related to sustainable housing, which is especially important outside the scope of building codes. Various kinds of systems or methods for sustainable housing, including assessment and rating systems, standards, and guidelines, are used in many countries of the world [5]. Moreover, energy conservation standards or energy consumption labeling systems for appliances and equipment are also used in many countries [6]. Such voluntary systems or methods usually include criteria, such as grades, levels, classes, target figures, or guideline values. Accordingly, referring to such existing voluntary systems, system designers can select suitable criteria to the desired values of variables. In addition, even if there are not exactly suitable criteria, referring to relevant systems and criteria usually provide system designers with information closely related to the variables and their desired values and help them determine the desired values.

In Tables 2 and 4 of Chapter 4, a high percentage of variables' desired values have been specified, based on this second approach. In particular, referring to the JHPIS and CASBEE for detached houses, we have selected appropriate grades or levels to the desired values of many variables. We have also used the long-life quality housing (LQH) certification criteria, when setting the desired values of several variables, including "total floor area." When determining the variable of "appliances" and its desired value, we have utilized the energy-saving labeling system of Japan. Furthermore, consulting several accessible and universal design guidelines has led us to specify the desired values of variables related to universal design, such as "grade of steepness" (slope).

In short, system designers in each country can compile the guidelines and checklist, referring to compulsory and voluntary systems related to buildings and housing used in that country.

2.4.2. *Application possibility to other categories*

Theoretically, the methodology can be applied to various categories of human activities. In other words, in the control system for promoting sustainable home design (Figure 5 in Chapter 4), "homes" in the block of "controlled objects" can be replaced with other categories of human activities [2]. Possibility of such replacement depends on if the table of relationships or the "sustainable design guidelines" can be compiled or not [2].

It will be probably easy to apply it to other types of buildings besides the home, because the structure is similar to one another [2]. It is also possible in theory to apply it to other kinds of infrastructure, including roads and parks [2]. Moreover, we consider it possible to apply the methodology to more large-scale and complex objects, including the city and town, for they are also considered as the complex of material and spatial elements [2].

2.5. Future research

2.5.1. Further case studies

The case study has supported the effectiveness of the methodology for promoting sustainable housing design. However, we are necessary to conduct further case studies, applying it to both new and existing houses. We expect that the increase of application cases also leads to an increase in the reliability of the methodology and help it to be widely used [2].

2.5.2. Revision of the guidelines and checklist

The "sustainable design guidelines" and "sustainability checklist" need to be updated, as occasion requires. Such occasion is projected to occur due to several causes, for instance, changes in the natural and social environment, developments in related sciences, innovations in related technologies, and response to the results of case studies [2]. Moreover, through such revision or update processes, we are going to investigate how to revise them efficiently.

3. Conclusion

This study has demonstrated the methodology for sustainable housing design by applying control science, with a case study. The main point of the methodology is the control system for promoting sustainable housing design with the sustainable design guidelines and sustainability checklist. Utilizing this methodology, we have actually designed a home and built it. The evaluations of the home indicate that closely following the methodology leads to comprehensively achieving sustainable homes with high environmental performance.

Meanwhile, we have pointed out several characteristics of the methodology, in addition to comprehensiveness. First, the diagram of the control system itself is beneficial because it concisely shows the whole picture of the sustainable design processes on both new and existing homes. Second, the "sustainable design guidelines" and "sustainability checklist" are user-friendly since the material and spatial elements are equivalent to real parts of homes. Moreover, the "element – variable – desired value" structure in the guidelines and checklist is superior in "adaptability to regional differences" and "flexibility toward changes over time."

In the twenty-first century, homes need to be transformed into those which contribute to deal with various issues, including climate change and financial problems due to aging population. Curbing the progress of climate change is a global challenge; therefore, mitigation measures have to be taken into homes all over the world. On the other hand, type and severity of impacts caused by climate change are different, depending on the region. Accordingly, appropriate adaptation measures need to be adopted in homes, in accordance with the predicted impacts in that region. Meanwhile, progressing aging population requires the inclusion of accessible and universal design into homes, in order to increase mobility of occupants and prevent injuries. Homes are used for a very long time; homes which are built or renovated now are expected to be used throughout the twenty-first century. Accordingly, such considerations toward sustainability need to be comprehensively taken into homes from the beginning.

Facing these circumstances, this user-friendly, comprehensive, adaptable, and flexible methodology is effective to promote sustainable housing design in various regions and countries. The guidelines and checklist shown in Tables 2 and 4 in Chapter 4 have been already compatible with climate change and aging population. These tables, which have been compiled to suit features in Japan, can be easily modified to fit features in other regions. Due to the same characteristic of this methodology, these tables can also be readily customized, so as to adapt to predicted impacts in each region caused by climate change. Hence, we expect that this methodology is used in various regions and countries, so as to facilitate sustainable home design.

The case study has successfully demonstrated the effects of the methodology on achieving sustainable homes. However, in order to confirm the effects, we need to conduct more case studies, applying it to both new and existing homes. Moreover, we will have to revise the "guidelines" and "checklist," as the occasion arises. Through such revision processes, we are planning to examine how to revise them efficiently. Meanwhile, it is theoretically possible to apply this methodology to other categories of human activities, which are regarded as the complex of material and spatial elements. We are also aiming to apply this methodology to more complex and larger scale human activities, such as the city and town.

Author details

Kazutoshi Fujihira

Address all correspondence to: fujihira@kankyogaku.com

Institute of Environmentology, Tokyo, Japan

References

[1] Bergman D. Sustainable Design: A Critical Guide. New York, NY, USA: Princeton Architectural Press; 2012. p. 144

[2] Fujihira K. System control for sustainability: Application to building design. In: Thomas C, editor. Complex Systems, Sustainability and Innovation. Rijeka, Croatia: InTech; 2016. pp. 37-55. DOI: http://dx.doi.org/10.5772/65875

[3] Gawande A. The Checklist Manifesto: How to Get Things Right. New York, NY, USA: Picador; 2009. p. 215

[4] Ching F, Winkel S. Building Codes Illustrated: A Guide to Understanding the 2015 International Building Code. Hoboken, NJ, USA: John Wiley & Sons; 2016. p. 448

[5] Carmody J, Weber W, Jacobson R. Center for Sustainable Building Research (CSBR) University of Minnesota. Design Guidelines for Sustainable Housing [Internet]. 2009. Available from: http://www.csbr.umn.edu/download/KoreaHousingProjectCombinedFinalReport.pdf [Accessed: 15-09-2017]

[6] The Institute of Energy Economics, Japan. Summary of Foreign Energy Conservation Measures Trends Survey (in Japanese) [Internet]. 2016. Available from: http://www.meti.go.jp/meti_lib/report/2016fy/000471.pdf [Accessed: 15-09-2017]